T0301326

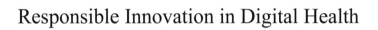

Responsible Innovation in Digital Health

NEW HORIZONS IN INNOVATION MANAGEMENT

Books in the New Horizons in Innovation Management series make a significant contribution to the development of Innovation Studies. As this field has expanded dramatically in recent years, the series will provide an invaluable forum for the publication of high-quality works of scholarship and show the diversity of issues and practices around the world.

Global in its approach, it includes some of the best theoretical and empirical work with contributions to fundamental principles, rigorous evaluations of existing concepts and competing theories, historical surveys and future visions.

Responsible Innovation in Digital Health

Empowering the Patient

Edited by

Tatiana Iakovleva

Professor of Entrepreneurship, UiS Business School, University of Stavanger, Norway

Elin M. Oftedal

Associate Professor of Leadership, Innovation and Market (LIM), Business Creation and Entrepreneurship, University of Tromsø, Norway

John Bessant

Professor of Innovation and Entrepreneurship, University of Exeter, UK, University of Erlangen-Nuremburg, Germany and University of Stavanger, Norway

NEW HORIZONS IN INNOVATION MANAGEMENT

Edward Elgar
PUBLISHING

Cheltenham, UK • Northampton, MA, USA

Published by
Edward Elgar Publishing Limited
The Lypiatts
15 Lansdown Road
Cheltenham
Glos GL50 2JA
UK

Edward Elgar Publishing, Inc.
William Pratt House
9 Dewey Court
Northampton
Massachusetts 01060
USA

A catalogue record for this book
is available from the British Library

Library of Congress Control Number: 2019938893

This book is available electronically in the **Elgar**online
Business subject collection
DOI 10.4337/9781788975063

ISBN 978 1 78897 505 6 (cased)
ISBN 978 1 78897 506 3 (eBook)

Printed and bound in Great Britain by TJ International Ltd, Padstow

Contents

Figures

Tables

Contributors

Allen Alexander is a Senior Lecturer in Innovation and Entrepreneurship at the University of Exeter Business School, Centre for the Circular Economy. He researches strategic knowledge management and how knowledge develops enhanced commercial capability and is a source of innovation. His recent studies have explored open innovation, innovation eco-systems and regional systems of innovation. He is working on exploring circular innovations, and their role in transitioning toward a circular economy. He has extensively explored the role of universities and academics in shaping corporate innovation and entrepreneurship practices.

Salomé Azevedo is a research (Innovation Management) and teaching assistant (Operations Management) at Católica-Lisbon School of Business and Economics. Her work focuses on how patients and caregivers develop treatments, therapies and medical devices to improve their life condition. She is Platform Manager at Patient Innovation. She did an internship at Carnegie Mellon University's Engineering and Public Policy Department. She holds a Master's degree in Biomedical Engineering from Instituto Superior Técnico, University of Lisbon.

Paul Benneworth is a Professor of Innovation and Regional Development at Høgskulen på Vestlandet, Bergen, Norway and a senior researcher at CHEPS, the Netherlands. Paul's research focuses on relationships between universities and societal development, notably affecting less powerful communities. He has written extensively on this topic, including the *University Engagement With Socially Excluded Communities* (2013, Springer), *Universities and Regional Development in the Periphery* (2018, Routledge) and *Universities and Regional Development: A Critical Assessment of Tensions and Challenges* (Routledge, 2012, with Romulo Pinheiro and Glenn C. Jones).

John Bessant, originally a chemical engineer, is currently Professor of Innovation and Entrepreneurship at the University of Exeter with visiting appointments at the universities of Erlangen-Nuremburg, Germany and Stavanger, Norway. In 2003 he was elected a Fellow of the British Academy of Management and in 2016 a Fellow of the International Society for Professional Innovation Management (ISPIM). He has consulted widely and is the author of 30 books and many articles on the topic. See www.johnbessant.org for more details.

Helena Canhão, MD, PhD is Full Professor at NOVA Medical School, Universidade Nova de Lisboa, Portugal. She is Head of EpiDoC Unit, CEDOC-NMS and Rheumatology Senior Consultant at CHLC – Hospital Curry Cabral, Lisbon. Graduate, PhD and Habilitation in Medicine from Lisbon School of Medicine, Universidade Lisboa. She holds a Master's degree in Medical Sciences from Harvard Medical School. She is Co-leader and Chief Medical Officer at Project Patient Innovation. President Elected of Portuguese Society of Rheumatology and Vice-President of the Portuguese League Against Rheumatism. She has authored several books, book chapters and scientific papers in peer-reviewed journals. See https://www.ncbi.nlm.nih .gov/pubmed/?term=canhao+h.

Lene Foss (Dr.oecon, PhD) is Professor of Innovation and Entrepreneurship at UiT – The Arctic University of Norway, School of Business and Economics. Her research concentrates on gender in innovation and entre-preneurship, academic/university entrepreneurship and entrepreneur-ship education. Her work appears in *Resources Policy, Small Business Economics, Journal of Small Business Management, European Journal of Marketing, International Small Business Journal, International Journal of Entrepreneurship and Behavioral Research, International Journal of Gender and Entrepreneurship, Education + Training, Industry and Higher Education* and *Education Research International.*

Verena Schulze Greiving has worked as a postdoctoral researcher in the Science, Technology and Policy Studies group (University of Twente, NL) on a project analyzing the notion of responsible research and innovation in eHealth applications. Additionally, she supports scientists to implement societal dimensions in their research both at the University of Twente and at Saxion University of Applied Sciences (NL). After a Bachelor's in Chemistry, she followed a Master's in Nanotechnology. She received her PhD from the BIOS Lab-on-a-chip group of Professor Albert van den Berg where she developed a microfluidic device for lipid bilayer experimentation.

Mark Griffiths is an Adjunct Professor in the Department of Finance and Business Economics in the Marshall School of Business at the University of Southern California. He was formerly the Jack Anderson Professor of Finance at Miami University's Farmer School of Business. Mark is widely published in leading entrepreneurship and management journals. He is a founding scholar of the *Entrepreneurship Research Society* and has won numerous teaching and research awards.

Tatiana Iakovleva holds a Professor of Entrepreneurship in Stavanger Business School, University of Stavanger, Norway. She received her PhD in Management from Bode Graduate School of Business, Norway, with focus on Entrepreneurship and Innovation. Her Master of Science in Business is from Norway (Bode) and from Russia (St. Petersburg). Dr. Iakovleva's research interests include personal and organizational antecedents leading to innovation and superior entrepreneurial performance on enterprise and regional level, female entrepreneurship, responsible innovation, as well as factors affecting entrepreneurial intentions.

Jill Kickul is a Professor in the Lloyd Greif Center for Entrepreneurial Studies and Research Director of the Brittingham Social Enterprise Lab at USC. She was formerly at New York University's Stern School of Business as Director of their Social Entrepreneurship Program. Jill is the Founding Director of The Annual Conference on Social Entrepreneurship and is the President-elect of United States Association for Small Business and Entrepreneurship. She has published more than 100 articles in leading entrepreneurship and management journals.

Kornelia Konrad is an Assistant Professor at the Science, Technology, and Policy Studies Department of the University of Twente. With a background in sociology, science, technology and innovation studies, her research interests focus on the role of anticipation in innovation processes, ranging from the study of hypes to the construction of socio-technical scenarios as an element of constructive technology assessment. In this vein, she is also involved in the development of approaches and tools for integrating considerations on societal impact and conditions into research and innovation, as a means towards responsible research and innovation.

Thomas Laudal is an Associate Professor of international business strategy at the University of Stavanger Business School. He received his PhD in Change Management (international textile/apparel industry) from the University of Stavanger. Laudal's research focuses on institutional and industry perspectives linked to the introduction of new ICT, new business models and sustainable business strategies. Laudal has more than 12 years of work experience from the European Commission, Norwegian ministries and from consultancy work in the oil and gas industry.

Bala Mulloth is an Assistant Professor of Public Policy at the Frank Batten School of Leadership and Public Policy at the University of Virginia. Mulloth's research areas include innovation and strategic processes within new ventures, sustainable business models, and social entrepreneurship. He was formerly a Visiting Faculty Fellow at National Defense University in Washington DC. He was also a faculty of Entrepreneurship at Central European University in Hungary. He holds a PhD in Technology Management from NYU Tandon School of Engineering.

Bernard Naughton is a postdoctoral researcher at the University of Oxford, Saïd Business School. His PhD concerned the management of digital drug screening technologies in the hospital setting. His current research concerns management and RRI (responsible research and innovation) in healthcare settings. Bernard is a clinical pharmacist by training and conducts impactful interdisciplinary research, bridging the gap between medical and social science. He also has expertise in the pharmaceutical supply chain, medicine falsification and associated regulations.

Elin M. Oftedal is an Associate Professor of Entrepreneurship in UiT School of Business and Economics, UiT – Norway's Arctic University. She received her PhD in Management from Bodø Graduate School. Her Master of Science in Business is from University of Stavanger and University of West Scotland. Dr. Oftedal participates in projects on sustainable technology such as renewable energy and carbon capture and usage (CCU), in addition to co-managing an international research project on responsible innovation in healthcare across six countries.

Pedro Oliveira is Professor MSO at Copenhagen Business School with special responsibilities in healthcare innovation, Academic Fellow at the Cornell Institute for Healthy Futures, Founder and President of Patient Innovation; and Co-founder of PPL-Crowdfunding. Previously he was a Professor at Católica-Lisbon School of Business and Economics, International Faculty Fellow at MIT-Sloan and Advisor to the Ministry of Science, Technology and Higher Education. He received his PhD in Operations, Technology and Innovation Management from the University of North Carolina at Chapel Hill; his MSc and 'licenciatura' from IST-Lisbon.

Luciana Maines da Silva is Assistant Professor at Unisinos University. Business degree from the São Judas Tadeu School of Accounting and Business, MBA from UNISINOS University and third-year PhD candidate in Business at UNISINOS University. Da Silva's research is focused on responsible innovation, dynamic capabilities, strategic management, emerging markets and in particular on stakeholder inclusion on the innovation process.

Andrea Marie Stangeland is a Service and Interaction Designer at Lyse. She has a Master's degree in Industrial Design from NTNU (Norwegian University of Science and Technology). She joined Lyse in 2015 as a Trainee in the Innovation department. Andrea has a great interest in the field of user experience and has been working on projects involving welfare technology and universal design.

Raj Kumar Thapa is a PhD scholar at University of Stavanger Business School, associated with the project 'Digitalize or Die – Dynamic Drivers of Responsible Research and Innovation in health and welfare services'. Raj holds Masters of Science in Innovation and Entrepreneurship from BI Norwegian Business School and his research interest is within responsible research and innovation, innovation and entrepreneurship and Social innovation.

Elisa Thomas works as a post-doctor at the Centre for Innovation Research at the University of Stavanger Business School, in Norway. Her research interests focus on entrepreneurship and innovation ecosystems, entrepreneurial university and open innovation intermediaries. She holds a PhD in Business Administration from Unisinos University, in Brazil, where she has worked as lecturer and course coordinator.

Marissa Titus is an Associate Consultant for Health Advocate's data analytics division, Engage2Health. She focuses on product design, project management, and business development. Marissa completed her Master's in Social Entrepreneurship at the USC Marshall School of Business and holds a Bachelor's in Molecular Biology with a minor in Health and Wellness from SUNY Binghamton. She is passionate about working at the intersection of these fields and is currently striving to improve health outcomes through health informatics.

Anna Trifilova is a Research Fellow at Exeter Business School, UK. She is also a Professor of Innovation at both Saint Petersburg University and Higher School of Economics in Russia. Anna is involved with ISPIM leading Teaching and Coaching Innovation SIG. At present, she is coordinating Erasmus+ TACIT project (Teaching and Coaching Innovation & Entrepreneurship Innovatively). She has been a part of research projects with a number of universities in Europe and with WWF, World Bank, and Moscow Government.

Dagfinn Wåge is Head of Innovation at Lyse Group. Dagfinn joined Lyse in 2006 coming from Telenor. He holds two master degrees (MSc and MoM) from Stavanger University and Norwegian business school BI respectively. He was Chief Product Officer and Chef Innovation Officer at Altibox before becoming Head of Innovation at the Lyse Group in 2012. Having a profound interest in the field of new digital business models, he recently published a book on this theme. See www.disruptiveecosystems.com.

Michael D. Williams, MD FACS is Associate Professor of Surgery and Public Health Sciences at University of Virginia. He serves as Director of the Center for Health Policy, and holds leadership roles as the Associate Chief Medical Officer for Clinical Integration, reflecting his interest in managing the health of populations. As his career has evolved, he has increased his understanding of the growing value of patient engagement in patient-centered care as the principal means of achieving a healthy, equitable society.

Danielle Wynne works as an innovation consultant at Bernoulli Analytics. Danielle holds a Master of Philosophy from the University of Exeter and a BSc in Geography from the University of Leeds. Danielle has eight years' experience in innovation and business strategy with a focus on geospatial analytics and GIS. She has worked on international projects with market-leading organisations covering a range of sectors including environmental reporting, engineering, aerospace, heritage, and health.

Acknowledgements

This book is a result of the four-year research effort by a multi-cultural research team from six different countries. Supported by the funding received from Norwegian Research Council, project number 247716/O70 'Digitalize or Die – Dynamic Drivers of Responsible Research and Innovation in Health and Welfare services', this work has truly influenced our lives. From being curious researchers we've moved to become engaged agents of change of the new phenomena we observed during this research. We are grateful for all contributors to this work, including our authors and informants. It would not be possible to write this book without inputs from entrepreneurs, patients, caregivers, healthcare professionals, researchers and policymakers, municipality and cluster representatives, innovation agents in healthcare sector that shared their stories and viewpoints with us. There are too many to mention by name but we're really grateful to all of them.

Special thanks are due to the team at Edward Elgar Publishing for their support during the writing and publication process.

1. Responsible innovation in digital health

Tatiana Iakovleva, Elin M. Oftedal and John Bessant

> There is only one way to look at things until someone shows us how to look at them
> with different eyes.
> Pablo Picasso

1.1 INTRODUCTION

There is little doubt that healthcare is in crisis. On the one hand there have been enormous advances in the nature and quality of care, several illnesses have been eradicated or minimized and lifespans have been extended to the point where some predict that the '100 year life' will be a realistic possibility for many children born today (Gratton and Scott, 2016). But, on the other hand many of these gains are becoming hard to sustain because of powerful forces on the demand side.

Healthcare costs and spending often rise at rates exceeding inflation, and are expected to increase in the future. For example, estimates suggests that aggregate healthcare spending in the United States will grow at an average annual rate of 5.8 per cent from 2015 through 2025 (Keehan et al., 2017). The health sector faces many challenges today and represents a significant cost of at least 10 per cent of GDP in the countries' economies. This situation is likely to worsen as a result of an aging population, rising prices and increasing complexity associated with technology in the health services (Marmot et al., 2012), where there is pressure on delivering qualitatively good services to all, while significantly reducing costs. Despite wide variations in healthcare funding systems the underlying trends are the same across countries.

Against this backdrop the need for disruptive innovation is clear and extensive efforts are being made to find a way out of the crisis through innovative new approaches. A major candidate here is 'digital healthcare' – an umbrella label for a wide range of technologies that could meet the healthcare challenges. Examples include various apps, telemedicine, electronic medical records, 'smart' homes and 'connected medicine'. The novelty of the field and the pace of underlying technological development means that both established companies and start-ups are engaging in this sector and empower patients to be a part of this development.

The technology allows for a reduction of costs due to remote processes and ability to treat a higher number of patients. Further, digital tools allow for the move to 'consumer-centric' healthcare, allowing citizens to take responsibility for managing their healthcare and that of their families. Their potential is significant – not just in terms of improving productivity within the healthcare delivery sector but also in offering better outcomes, higher quality and reliability, greater patient autonomy and higher quality of life.

Taken at face value digital healthcare appears to offer a rosy future for patient-centred high quality healthcare delivery at an affordable cost and open to all. But, the promise of digitalization might not be attained unless we are aware of its challenges. What if these miraculous technological developments are abused, exposing society to the 'dark side' of digital innovations? For example, electronic records and big data diminish privacy, and could lead to screening and refusal of care. Intelligent homes may not be designed around patients, but for the convenience of the 'system' and give the patient an experience more akin to prison than home or hospital, monitored and managed by robots and sensors. Automated systems might increasingly make their own decisions about treatment and even life and death.

This raises an old question – the role of responsibility in decisions about technological innovation. It is timely to shift attention from 'how to achieve ground-breaking innovation' towards 'how to achieve ground-breaking innovation in a responsible manner'.

1.2   RESPONSIBLE INNOVATION

For more than fifty years researchers have been arguing for approaches which involve some degree of technology assessment and control, focusing on tools that allow anticipation and exploration of likely consequences of technology decisions. Discussions around this theme go back at least to the 1970s with the 'Science, technology and society' (STS) movement and the establishment of key influential organizations such as the Science Policy Research Unit at the University of Sussex (Cole et al., 1973). Its most recent manifestation can be seen in work on 'responsible innovation' (Owen, Bessant and Heintz, 2013). 'Responsible innovation' (RI) can be defined as 'a transparent, interactive process by which societal actors and innovators become mutually responsive to each other with a view to the (ethical) acceptability, sustainability and societal desirability of the innovation process and its marketable products (in order to allow a proper embedding of scientific and technological advances in our society)' (von Schomberg, 2011, p.9).

Thus, the focus of RI is on considering alternative outcomes and engagement of multiple stakeholders early in the innovation process. A helpful framework focused on four dimensions of responsible innovation is offered by Stilgoe, Owen and Macnaghten (2013) which looks at anticipation of risks, user inclusiveness, reflexivity and responsiveness.

In outline, these cover questions like:

Have the consequences of implementing the innovation been explored? What alternative scenarios might there be? (anticipation);

Have the 'owners' of the business model been sufficiently reflective in exploring and developing it – or does it represent a single perspective or an implicit dominant design? (reflexivity);

Does the model take into account the inputs of relevant stakeholders? (inclusivity) and

Can the innovation be adapted in response to the answers to these questions, is there flexibility in delivery? (responsiveness)

1.3   USER INCLUSION – THE MISSING LINK IN HEALTHCARE INNOVATION?

The third dimension is of particular relevance in our healthcare context. In the process of organizing healthcare as a *system*, trying to meet the varied and complex needs of many different individuals there is a risk that the connection to the end-user, the patient, gets lost. The increasing professionalization of healthcare and the rising sophistication of the technology involved has meant that the 'voice of the patient' has often been drowned out. Instead, patient input to the innovation process is being replaced with a passive status which sees them as recipients of healthcare rather than able to shape it.

There is growing recognition of this divide and the rhetoric within many healthcare systems is now around bringing the patient back into the equation. There are several reasons why this might be a timely move; first there is good evidence that involving patients leads to better design because they understand their needs best. Research suggests that patient-led innovation has a strong track record of success, not least because articulating tacit knowledge about the patient experience and then building that into design of new solutions can enrich those solutions and make them compatible with a wider downstream set of users (Oliveira et al., 2018). Increasingly patients are being seen not as passive consumers of healthcare but as active partners in its creation and delivery.

Considering the user dimension also helps us explore a paradox associated with healthcare technology. As Hwang and Christensen (2008), argue, despite so many sophisticated medical technologies introduced every year, healthcare has not been disrupted to a significant degree. Their view is that this happens because technology has almost always been implemented in a *sustaining* manner in healthcare – primarily to help hospitals and doctors to solve the most complex problems. As a consequence most technological enablers have failed to bring about lower costs, higher quality and greater accessibility (Christensen, Waldeck and Fogg, 2017).

The theory of disruptive innovation suggests that new trajectories emerge at the edges of the mainstream, working with users who are not served or under-served in the current provision. Arguably, bringing this group of stakeholders into the development equation might open up the possibilities for significant shifts in healthcare performance, enabling radical changes of the kind which are urgently needed.

The potential for significant change through user involvement is there; the question is to what extent is it, or can it be realized? One risk with the powerful new technologies we list above is that patients might become further margin-alized – healthcare is something done to them and delivered in an increasingly non-human fashion. An alternative view is to put the patient at the centre and explore models which would empower them to shape and direct the technolo-gies in their own interests. And that begs a second question – ***how to empower users to become a part of innovation process?***

1.4   OVERVIEW OF THE BOOK

This book tries to explore the question of responsible innovation in the emerging field of digital healthcare and focuses particularly on the challenge of increasing user engagement in the process. It draws on a variety of case examples of innovations in digital health from six different countries – Brazil, the Netherlands, Norway, Portugal, UK and USA.[1]

In Chapter 2, we introduce the theoretical construct of responsibility and its implications for the firm-level innovation process. Chapter 3 discusses the context for digital healthcare and compares different healthcare systems, analysing the power centres and the role of patients in such systems. We suggest that patients are not a homogeneous group in terms of their innovation behaviour. Instead there is a spectrum of behaviours – from passive recipients of healthcare services through to the 'informed' patient, who is equipped to use technology based on improved understanding. Beyond this, and enabled by digital technology, the 'involved patient' can play an active role within a wider healthcare delivery system by actively providing feedback to it. At the very extreme of this spectrum we find the 'innovative patient', who supplies ideas of their own based on their deep understanding of their healthcare issues.

Chapter 4 to 14 of this book provide empirical examples of digital healthcare innovations and discusses four dimensions of RI in each of them. Chapter 4 written by Pedro Oliveira, Salomé Azevedo and Helena Canhão illustrates that patients all over the world can be real innovators and entrepreneurs. The social media platform 'Patient Innovation.com' described in this chapter gathered over 800 innovative solutions developed by patients from over 70 different countries. The platform enables patients to share their solutions and help innovation diffusion.

Chapter 5 by Thomas Laudal and Tatiana Iakovleva discusses how patient feedback can potentially lead to service innovations in hospitals, looking at the example of a Norwegian hospital. It also debates how the introduction of digital communication channels like electronic health records (EHR) can trigger this process. The authors identify three questions for future research: What kind of relationship is there between the release of EHRs and patient feedback? How do patient-initiated innovations influence hospital performance? And, what are the most important contextual factors?

Chapter 6 written by Bala Mulloth and Michael D. Williams illustrates how patients can become more informed and involved in their care by describing the University of Virginia (USA) digital healthcare application. The authors analyse the process through the lens of responsible innovation and present the challenges to realizing the full potential of this type of EHR.

Chapter 7 by John Bessant, Allen Alexander, Danielle Wynne and Anna Trifilova explores the theme of 'design space' around digital innovations. Using the particular example of a detailed longitudinal case of the development and diffusion of a digital health information platform the authors identify a number of key points at which the innovation concept 'pivoted' to reflect new information, some of which resulted from a wider level of inclusion.

In Chapter 8 Raj Kumar Thapa and Tatiana Iakovleva addresses the question of how business organizations pursue responsible innovation in business development and create positive social impact. Based on an explorative case study of a privately owned Norwegian firm within a medical industry, this chapter analyses purpose, process, and outcomes of the innovation process on the firm level from the responsibility point of view.

Chapter 9 written by Dagfinn Wåge and Andrea Marie Stangeland describes the development of the application of a video communication channel in homes of elderly people by a medium-sized Norwegian company. Presented from a firm perspective, this story provides an example of the challenges of the new technology development in this particular setting and learnings gained, while balancing new technology development with needs and feedbacks from users as well as market considerations.

In Chapter 10 Elin M. Oftedal and Lene Foss discuss how responsible start-ups are dealt with by the health sector. Through following three Norwegian companies, the authors acquire insight into challenges the entrepreneurs experience when they introduce their technology/service to the healthcare sector.

Chapter 11 by Kornelia Konrad, Verena Schulze Greiving and Paul Benneworth describes the innovation process of an eHealth application which emerged as a user-driven, local project in the Netherlands. This chapter considers if and how the regional and partly local, bottom-up nature of the innovation network was conducive to enacting dimensions of responsibility.

In Chapter 12 Elisa Thomas and Luciana Maines da Silva analyse how two Brazilian digital health start-ups manage stakeholders' participation. Results show that the inclusion of stakeholders happens via tacit or explicit knowledge exchange, and that firms need different structures and routines to deal with different knowledge types to fully explore the potentials of inclusiveness.

Chapter 13 written by Bernard Naughton and Lene Foss explores how an academic entrepreneur pursues responsibility in commercializing research-based knowledge. Through a narrative based case study of a university professor employed at a mid-sized UK university, the authors provide insight into how a professor has performed digitalized healthcare practice research, innovation and commercialization.

Chapter 14 by Jill Kickul, Mark Griffiths and Marissa Titus investigates a digital therapeutics case from USA. They identify two major challenges for acceptance of digital therapeutics: proof of efficacy and consumer habits.

Our final chapter, Chapter 15, summarizes the need for responsible innovation in digital health. Far from being a burdensome requirement our research suggests that including users actively and early on in the design process can make a significant difference both to the quality of innovation and to its downstream acceptance. It allows early identification of unnecessary risks in implementing innovations. Through feedback from users and stakeholders designers can adapt and adjust business models, pivoting them to better fit the emerging context and meet the needs of diverse stakeholders. For policymakers the nature of the healthcare system and the ways in which procurement often take place offer a significant opportunity to shape innovation towards responsible trajectories which put the concerns of patients at the forefront.

We have tried in this book to offer evidence and examples of user inclusiveness, early anticipation of risks, reflectiveness of business models in digital innovations in health and welfare sectors. We observe these phenomena in different situations and variety, and within different international context. Whether we are speaking about user-driven innovations, or start-up sustainable models, or pathways to responsible innovation in mature organization, the same important issues of inclusivity, anticipation and responsiveness arise. We believe that a responsible innovation process can be a tool for users, clinicians, businesses and policymakers to create a system where the newest technology can be implemented to increase care for the patient and minimize cost for the care-provider.

NOTE

1. Empirical evidence comes from project 'Digitalize or Die – Dynamic Drivers of Responsible Research and Innovation in Health and Welfare services', funded by Norwegian Research Council, project number 247716/O70.

REFERENCES

Christensen, C., Waldeck, A. & Fogg, R. (2017). How disruptive innovation can finally revolutionize healthcare. Innosight executive briefing, retrieved on 8 March 2019 from https://www.christenseninstitute.org/wp-content/uploads/2017/05/How-Disruption-Can-Finally-Revolutionize-Healthcare-final.pdf.

Cole, H. S. D., Freeman, C., Jahoda, M. & Pavitt, K. L. R. (1973). *Thinking about the Future: A Critique of 'Limits to Growth'.* Published on behalf of Sussex University Press, London.

Gratton, L. & Scott, A. (2016). *The 100-Year Life: Living and Working in an Age of Longevity.* London: Bloomsbury Publishing.

Hwang, J. & Christensen, C. M. (2008). Disruptive innovation in health care delivery: a framework for business-model innovation. *Health Affairs*, 27(5), 1329–1335.

Keehan, S. P., Stone, D. A., Poisal, J. A., Cuckler, G. A., Sisko, A. M., Smith, S. D. ... & Lizonitz, J. M. (2017). National health expenditure projections, 2016–25: price increases, aging push sector to 20 percent of economy. *Health Affairs*, 36(3), 553–563.

Marmot, M., Allen, J., Bell, R., Bloomer, E. & Goldblatt, P. (2012). WHO European review of social determinants of health and the health divide. *The Lancet*, 380(9846), 1011–1029.

Oliveira, P., Zejnilovic, L., Azevedo, S., Rodrigues, A. M. & Canhão, H. (2018). Peer-adoption and development of health innovations by patients – a national representative study of 6204 citizens, *JMIR Preprints*. 30/07/2018:11726, doi: 10.2196/preprints.11726, http://preprints.jmir.org/preprint/11726.

Owen, R., Bessant, J. & Heintz, M. (eds) (2013). *Responsible Innovation: Managing the Responsible Emergence of Science and Innovation in Society.* Chichester: John Wiley & Sons.

Stilgoe, J., Owen, R. & Macnaghten, P. (2013). Developing a framework for responsible innovation. *Research Policy*, 42(9), 1568–1580.

von Schomberg, R. (2011). Towards responsible research and innovation in the information and communication technologies and security technologies fields. A Report from the European Commission Services, Directorate General for Research and Innovation, Luxembourg: Publications Office of the European Union, 2011, retrieved on 8 March from http://ec.europa.eu/research/science-society/document_library/pdf_06/mep-rapport-2011_en.pdf.

2. Responsible innovation as a catalyst of the firm innovation process

Tatiana Iakovleva, Elin M. Oftedal and John Bessant

2.1 THE PROMISE OF RESPONSIBLE INNOVATION

While technology keeps developing, the ethical problems of utilizing it for social purposes have been debated for an extended period of time. Originating in post-war concerns about the potential negative impacts of technologies like nuclear power the Science, Technology and Society (STS) community emerged and became an influential force in discussions from the 1960s onwards. Groups like the Science Policy Research Unit at the University of Sussex in the UK were established to concentrate research in this field and an extensive literature emerged from these roots around technology assessment and the tools and techniques for operationalizing it (Cole, 1973; Freeman, 1984).

Whether it deals with genetic engineering, nanotechnologies or digitalization, responsible innovation (RRI) has a long heritage as a field of research and practice. Today the discussion is carried on around key themes such as sustainability, ethics and social responsibility in a wide range of books and journals (Owen, Bessant and Heintz, 2013; Stilgoe, Owen and Macnaghten, 2013). Its most recent manifestation can be seen in work on 'responsible innovation' (Owen, Bessant and Heintz, 2013).

As revealed by a recent literature review, over five hundred articles have been published on the topic of Responsible Research and Innovation (RRI) between 2003 and 2016 across 208 different journals (Thapa, Iakovleva and Foss, under revisions). While RRI often looks at the scientific aspect and the development process in 'grand challenges' like climate change, resource depletion, poverty alleviation, ageing societies and so on, RI has a more fine-grained focus on the innovation itself (cf. von Schomberg 2013, Blok and Lemmens, 2015). In this chapter we focus our attention on RI. 'Responsible innovation ' (RI) can be defined as 'a transparent, interactive process by which societal actors and innovators become mutually responsive to each other with a view to the (ethical) acceptability, sustainability and societal desirability of the innovation process and its marketable products (in order to allow a proper embedding of scientific and technological advances in our society)' (von Schomberg, 2011, p. 9).

There has been extensive concept development in responsible research and innovation (RRI) (Genus and Stirling, 2018; Stilgoe, Owen and Macnaghten, 2013; Owen et al., 2013; Ribeiro et al., 2018) but these discussions are not yet concentrated into a particular field; instead RI is a truly cross-discipline debate. Since 2013 the number of publications on this topic has increased dramatically and continues to grow. The majority of RRI research is concentrated in and around sensitive areas of technological innovation such as nanotechnology, biotechnology, gene-driven technology, digital technology (Thapa, Iakovleva and Foss, under revisions). Recent analysis of 126 conceptual papers revealed that research in this field can be broadly categorized into four domain ontologies as follows: RRI drivers, RRI tools, RRI outcomes and RRI barriers (Thapa, Iakovleva and Foss, under revisions). For simplicity, we use term RI rather than RRI in further discussions.

Despite the growing body of literature, the concept of RI is still largely normative in nature. In this chapter we argue that the concept of RI is an emerging one, and not fully conceptually stable, having a number of characteristics which are not worked through. RI's basis is that 'responsibility' in an innovation emerges when societal actors have opportunities to endorse behaviour that influence the innovative process in translating an idea into a launched product, service or technique. However, questions still remain about which stakeholders to include, at what stage and under what circumstances? The effectiveness of an inclusion exercise can be determined by how efficiently, complete and relevant information is obtained from all appropriate sources, transferred to and processed by those responsible and combined to generate a response (Rowe and Frewer, 2005). Nevertheless, over-inclusion of participants in the decision-making process, on many occasions, could jeopardize the process through compromising integrity (Spinello, 2003) and through information asymmetry (Blok, Hoffmans and Wubben, 2015). Further, there is a danger that the RI trajectory might not only encounter obstacles but might also be directed against the RI aspiration and be perceived as a development barrier in traditional corporate cultures (Grinbaum, 2013). As such the promise of RI could transfer to a challenge to innovative behaviour. Therefore some clues should be developed as to which stakeholders to include.

One might also question whether reflection is necessary for the institutional circumstances and the characteristics of the innovation system as enabling or constraining the suggested RI dimensions. This concern for understanding the conditions for RI also emerges in Walhout and Konrad (2015). There are many existing 'de facto' practices, processes and governance arrangements on existing actor landscapes that might affect RI in the making (Walhout and Konrad, 2015, p. 48). This calls for more attention of different kinds of contexts where 'responsibility in the making' is evident.

In this chapter we first provide an overview of the field, followed by a discussion of purpose, process and outcome of innovation. Innovation is described as a complex process in the context of uncertainty where design space occupies an important role. We conclude by arguing that at the high levels of uncertainty involved in radical or disruptive innovations, there is a need to keep design space open for deliberate inclusion allowing anticipation and reflexivity to happen.

2.2 TOWARDS A CONCEPTUAL FRAMEWORK OF RESPONSIBLE INNOVATION ON THE FIRM LEVEL

The development of novel solutions is a complex process, a journey which requires the management of several factors at different stages – concept, design, testing, launch (Tidd and Bessant, 2014). Although user input would benefit all stages, from initial ideation, through various test and refine cycles to final launch and diffusion, analysis of empirical studies of innovation projects from a RI perspective shows that inclusion often happens only on the late test stage of the innovation process (Silva et al., forthcoming). This is often a stage which only allows for minor adjustments of the solution; any serious changes are likely to have significant cost and/or delay implications for the project.

RI drivers focuses on the antecedents of responsible innovation. Drivers relates to the basic conditions necessary for responsibility in innovation process to emerge. What drives RI is engagement, in particular it is the purpose of the innovation to tackle societal challenges. This is related to the engagement of users, customers, relevant stakeholders, experts, policymakers, politicians and the public in the early stage of the research and innovation process by way of active and deliberate participation. Drivers includes pre-engagements (te Kulve and Rip, 2011); stakeholder engagement (Allon et al., 2016; Gudowsky and Peissl, 2016; Schroeder et al., 2016), upstream and public engagement (Bronson, 2015; Vincent, 2013), and the transdisciplinary approach (Rose, 2012; Stilgoe, 2012; Pierce, 2013). Although different themes appear within this domain, the bottom line is the inclusion of different actors in research and innovation activities. The inclusion of different actors in the innovation process ensures that they contribute with their knowledge, experience, challenges, expertise and opinions.

Organizations might start with a lack of awareness of RI and then over time develop increasing awareness and use RI as a deliberate strategy, as suggested in the maturity model (Stahl et al., 2017). RI can be considered as a tool for extracting and exploiting the best knowledge for innovation and for shaping research and innovation towards desirable innovation outcomes which are, socially, economically, and environmentally robust. To achieve this, purpose, process and outcome of innovation should be considered from the responsibility point of view (Stilgoe, 2012; Stahl et al., 2017).

'Purpose' refers to the reason(s) behind the innovation or the idea in focus and is strongly related to the motivation behind the activity. The purpose of responsible innovation should be emphasized based on the needs of society, for instance alleviating societal problems and keeping in mind to align the innovation activities as per the expectation of society and not just avoiding harm.

'Process' refers to all of the activities that are undertaken in the pursuit of responsibility. Process might include such elements as anticipation, reflection, inclusion and responsiveness.

Finally, 'outcomes' lead to certain identifiable consequences. It is to be acknowledged that RI processes will not necessarily bring responsible outcomes, but help develop a culture of lifecycle thinking (Köhler, 2013), reflecting and responding to take care and protect the society and environment from further damage (Deblonde, 2015). The aspiration of RI can be achieved only if organizations demonstrate responsiveness, meaning either abandoning manufacturing products which add negative externalities in society and to the environment or reviewing and redesigning them to provide an alternative solution. This could enhance trustworthiness of the business organizations to society, which could in turn improve reputation and brand image. Furthermore, a responsible attitude would build individual and collective capability to direct research and innovation towards the socioeconomic transformation of society (Voegtlin and Scherer, 2015). We summarize purpose, process and outcome of RRI in Figure 2.1.

Innovation process theories typically describe a 'development funnel' as a sequential process consisting of several stages, including outlined concept, detailed design, testing and launch phases (Tidd and Bessant, 2014). Although recent literature has extended far beyond the classical 'technology-push' model of innovation (Rothwell, 1994) and advanced our view on innovation process through more flexible innovation models (MacCormack and Verganti 2003), still a lot of innovations are locked into the 'dominant design' of the solution too early after the concept phase. With low levels of market and project uncertainty such lock ins provide higher efficiency of the innovation process. In the context when there are higher levels of uncertainty both for solution and for the market, there is evidence that there is a need to hold the design space open for a much longer time to allow reflections and modification of the solution before the dominant design is chosen (MacCormack and Verganti, 2003).

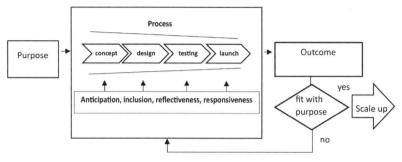

Source: Authors.

Figure 2.1   *Responsible innovation process for the economic actors*

To help organizations to implement RI in their practices, it is useful to talk about tools identified earlier in the literature. RI tools include methods or approaches intended for the effective engagement, foresight, anticipation and mitigation of potential risks that novel technology innovation might bear. This is to ensure that the particular innovation is aligned with its purpose norms in addition to values and expectations of society. It includes such tools as: walk-shop approach (Wickson and Miller 2015); engagement workshops (Selin et al., 2015; Stahl and Coeckelbergh, 2016; Rerimassie, 2016), online platforms (van Oost et al., 2016); online knowledge sharing (Vogel, 2012), social exper-imentation and design thinking (Stilgoe, 2012; Stilgoe, 2016), anticipation of risks and technology assessment (Le Feuvre et al., 2016; Inghelbrecht et al., 2016) as well as other techniques. Knowledge, in this context, plays a crucial role. Innovators need to acknowledge the fact that knowledge within a par-ticular community would still be limited to address overall socioeconomic, environmental and ethical issues in society. The themes within the RI tools domain are therefore highly concentrated on possible ways of accumulating knowledge and successfully deploying it in order to overcome societal and environmental challenges together.

RI barriers focus on the potential hindrances that RI practice may face. Directing innovation towards 'societal desirability' could be challenging. Consequently, the themes within this book outline obstacles and benefits that may arise while implementing RI. For example, RI promotes open access to research and innovation results (Rose, 2012; Gupta et al., 2016). However, the predominant practices of protecting intellectual property rights through patents, copyrights and trademarks could challenge RI and its successful tran-sition as the debate on the relevance of protecting intellectual property rights in research and innovation is an ongoing one (Spinello, 2003). As discussions in this domain revealed, research and innovation seem to be highly influenced and governed by the interests of certain groups of actors. For instance, in pursuit of gaining quickly competitive advantage, businesses might demonstrate ignorance of ethical and environmental issues, either intentionally or unintentionally (Blok, 2016). By contrast, societal and environmental activists oppose such practices and force businesses to abandon them. Bringing all these conflicting interest groups together and negotiating about so-called 'societal desirability' as suggested by RI might cause a conflict of interests (Taddeo, 2016). Creating harmonious or standardized RI on a global scale could encounter obstacles due to multiple values, interests and perceptions of what is 'responsible' or 'irresponsible' research and innovation (Arnaldi and Gorgoni 2015; Ruggiu, 2015).

2.3 FOUR DIMENSIONS OF RESPONSIBLE INNOVATION PROCESS

As we can see from discussion above, the theoretical development has so far focused on making normative models of responsibility. As such, the term responsibility could be critiqued for not being specific enough and thus complicate the already complex concept of research and innovation. Stahl and Coeckelberg (2016) recommend studying the issue of responsibility for the individual actor in industries. Therefore it would be useful to investigate how industrial actors approach responsibility and whether it is in line with the theoretical development.

Extending the RI model suggested by Stahl et al. (2017) we argue that for innovation to be able to diffuse in a responsible way, its purpose, process and outcome with regard to ethical and responsible behaviour should consider four elements suggested by Stilgoe, Owen and Macnaghten (2013): anticipation, inclusiveness, reflectiveness and responsiveness. In outline, these dimensions might be described in the following way:

- Anticipation – describing and analysing those intended and potentially unintended impacts that might arise, be these economic, social, environmental, or otherwise.
- Reflection – reflecting on underlying purposes, motivations, and potential impacts, and on associated uncertainties, risks, areas of ignorance, assumptions, questions, and dilemmas.
- Inclusivity – opening up visions, purposes, questions, and dilemmas to broad, collective deliberation through processes of dialogue, engagement, and debate, inviting and listening to wider perspectives from public and diverse stakeholders.

• Responsiveness – using this collective process of reflexivity to both set the direction and influence the subsequent trajectory and pace of innovation.

Below we broadly discuss these constructs.

1. Inclusion refers to the involvement of different stakeholders in innovation activities in order to represent their ideas, creativity and voices. The inclusion of public and all the relevant actors in the governance of science and innovation is a growing requirement for legitimacy (e.g., Irwin, 2006; Hajer, 2009). This opens the platform for dialogue and discussion that provides social intelligence, which would mediate in avoiding adverse public relations (Stirling, 2006). Furthermore, inclusion and deliberative participation of different actors in the innovation process helps the development of perceived ownership of the innovation outcomes and motivates creativity (Ayuso, Rodríguez and Ricart, 2006).

2. Anticipation is systematic thinking about emerging critical issues and discovering new possibilities and opportunities (Martin 2010; Stilgoe Owen and Macnaghten, 2013). This increases socially desirable outcomes, by shaping the future and organizing the essential resources for desirable innovation outcomes (te Kulve and Rip, 2011). To increase the probability of a highly positive social impact of innovation outcomes, anticipation in the very early stage of innovation activities is necessary (Owen et al., 2013). In this context anticipation demands focus beyond traditional risk–benefit analysis in terms of profit/loss (von Schomberg, 2011).

3. According to Stilgoe, Owen and Macnaghten (2013) reflexivity in the context of RI involves holding a mirror up to one's activities commitments and assumptions, being aware of the limits of knowledge and being mindful that a particular framing of an issue may not be universally held (Stilgoe, Owen and Macnaghten, 2013, p. 1571). RI demands for reflexivity in relevant actors in order for them to critically assess their own preconceptions. Reflexivity calls for firms to play close attention to value systems and social practices in the innovation process. The more reflexive the firm, the more trustworthy they will be in the public sphere. Reflexivity is vital in building a stronger and sustainable alliance, partnership, or collaboration among different internal and external knowledge networks, and image and reputation among the customers and users (Stahl et al., 2013). This will then assist in retaining old customers and increasing new ones.

4. Responsiveness assures the relevant actors that their ideas, views, or voices are taken into consideration since it is about demonstrating a deliberative attitude (Meijboom, Visak and Brom, 2006). It is about flexibility of innovation trajectories. Responsiveness ensures the ability to show care and respect towards stakeholders and societal values. This is directed

towards minimizing the blame game among stakeholders, collaborators, partners and other key stakeholders in the innovation process and towards co-responsibility.

Implementation of responsible innovation on the firm or project levels requires a good fit between the purpose that aims at innovating for the benefits of society and outcomes of innovation. In order to achieve this fit, inclusiveness seems to be a key necessary element, which ultimately leads to better anticipation. Further, adaptive capabilities are necessary to be able to reflect on feedback and respond to demanded changes in a responsible way. Thus, the process of participation and deliberate inclusion of relevant stakeholders becomes crucial during the whole innovation process.

Involving stakeholders makes decision-making processes more open and participatory, more focused on sustainable development. The search for solutions demands the inclusion of stakeholders that opens up for new visions, purposes, questions and dilemmas. This broadens the collective deliberation through processes of dialogue, engagement, and debate, inviting and listening to broader perspectives of audiences and diverse stakeholders, and revolves around a quest for social legitimacy for innovation (Demers-Payette, Lehoux and Daudelin, 2016). The main goal of inclusion is to diminish the authority of experts, with the inclusion of new voices in the governance of science and innovation as part of a quest for legitimacy. (Lubberink et al., 2017). This allows the introduction of a wide range of perspectives to reformulate issues and to identify potential contestation areas (Owen et al., 2013).

However, stakeholders' inclusion into the innovation process may also cause some challenges. Stakeholders might hold quite different and often conflicting views on the problem in question, based on their diverse political, cultural economic and social embeddedness (Kuzma and Roberts, 2018). Another critical point is that excessive inclusion may jeopardize the integrity of the common good (Spinello, 2003), as well as informational asymmetry (Blok, Hoffmans and Wubben, 2015). Such conflicts can lead to slowing down the decision process (te Kulve and Rip, 2011), and consequently, innovation process. Thus, for the economic agent, it becomes of crucial importance to recognize that inclusion at an early stage is a necessary condition to later acceptance and diffusion of innovation, but that this might also slow down the innovation process.

The question arises around the conditions under which inclusion is a necessary and useful tool. Moreover, who should be relevant stakeholders, and to what degree should they be involved in the innovation process – should they just provide inputs, or have a voice during the decision-making process? Lean models of innovation and agile methods emphasize user inclusion at early stages and continuous interactions during the whole process (Ries,

2011; Cooper, 2017). The key characteristic of such methods is interactivity prolonged design space, which allow to 'fail cheaply' and find alternative futures for solutions. Florén et al. (2017) argue that innovation project failures often occur due to too early lock-ins in the 'fuzzy front end' of innovation, when dominant designs were established too early without enough inputs or experimentation. Thus, RI principles of anticipation, inclusion and reflection fit well and add to well-accepted methods that allow flexibility and prolonged design space for innovation processes. Such methods have proved to be especially valuable under conditions of high uncertainty – both technological and in relation to market. Such high uncertainty is often associated with radical and disrupting innovations (Christensen, Raynor and McDonald, 2015) and this suggests that the principles of RI can be useful in this context in particular.

2.4   FUTURE RESEARCH AGENDAS

Evidence suggests that innovations which are pushed to the market without adequate consideration of their social desirability or their fit with underlying values are often rejected (Owen et al., 2013). It therefore makes sense to carefully consider incorporating users as early and as extensively as possible within the innovation process. Doing so requires thinking about several key questions, particularly around who to include, how to include them and when. In this chapter we described responsible innovation from a firm perspective, emphasizing that purpose, process and outcome of innovation should all be done in a responsible way to ensure higher innovation diffusion. We outlined anticipation, inclusion, reflection and responsiveness as main elements of the process dimension, as all four dimensions are crucial for making innovation flexible and adjustable to achieve the desired outcomes.

We also need to recognize that innovations do not happen in a vacuum. They are embedded in particular contexts – country or region – and we need to take the influence of these into account. Responsible innovation raises questions of broader inclusion, not only of local stakeholders, but also a wider set of views and perspectives, for example around gender or race inequality and about inequality of knowledge (Evers, 2002). This raises an even broader question of how to make innovative solutions responsible across countries, for societies with quite different knowledge bases and value systems? Would inclusion of only local stakeholders into the innovation process ensure globally responsible outcomes?

Disruptive innovations, like those in the field of digital healthcare have the potential to change and challenge established systems, and so it is important to ensure they are designed and diffused in a responsible way. In this chapter we have tried to show that principles of RI are not something that is required out of philanthropy, but rather that exercising responsible innovation might enhance successful commercialization. This provides a hope that recognition of the inclusion, anticipation and reflection principles will become a natural part of accepted innovation practice by economic actors.

REFERENCES

Allon, I., Ben-Yehudah, A., Dekel, R., Solbakk, J.-H., Weltring, K.-M. & Siegal, G. (2016). Ethical issues in nanomedicine: tempest in a teapot? *Medicine, Health Care and Philosophy*, 1–9. doi: 10.1007/s11019-016-9720-7.

Arnaldi, S. & Gorgoni, G. (2015). Turning the tide or surfing the wave? Responsible Research and Innovation, fundamental rights and neoliberal virtues. *Life Sciences, Society and Policy*, 12(1), 6. doi: 10.1186/s40504-016-0038-2.

Ayuso, S., Rodríguez, M. A. & Ricart, J. E. (2006). Using stakeholder dialogue as a source for new ideas: a dynamic capability underlying sustainable innovation. *Corporate Governance: The International Journal of Business in Society*, 6(4), 475–490.

Blok, V. (2016). Bridging the gap between individual and corporate responsible behaviour: toward a performative concept of corporate codes. *Philosophy of Management*, 1–20. doi: 10.1007/s40926-016-0045-7.

Blok, V. & Lemmens, P. (2015). The emerging concept of responsible innovation. Three reasons why it is questionable and calls for a radical transformation of the concept of innovation. In Koops, B-J., Oosterlaken, I., Romjin, H., Swierstra, T., and van den Hoven, J. (eds), *Responsible Innovation 2*. Cham: Springer, pp. 19–35.

Blok, V., Hoffmans, L. & Wubben, E. F. M. (2015). Stakeholder engagement for responsible innovation in the private sector: critical issues and management practices. *Journal on Chain and Network Science*, 15(2), 147–164.

Bronson, K. (2015). Responsible to whom? Seed innovations and the corporatization of agriculture. *Journal of Responsible Innovation*, 2(1), 62–77.

Christensen, C. M., Raynor, M. E. & McDonald, R. (2015). What is disruptive innovation? *Harvard Business Review*, 93(12), 44–53.

Coeckelbergh, M.(2016). Technology and the good society: a polemical essay on social ontology, political principles, and responsibility for technology. *Technology in Society*, 1–6. doi:http://dx.doi.org/10.1016/j.techsoc.2016.12.002.

Cole, H. S. (1973). *Thinking about the Future: A Critique of the Limits to Growth*. London: Chatto & Windus.

Cooper, R. (2017). *Target Costing and Value Engineering*. London: Routledge.

Deblonde, M. (2015). Responsible research and innovation: building knowledge arenas for glocal sustainability research. *Journal of Responsible Innovation*, 2 (1), 20–38. doi: 10.1080/23299460.2014.1001235.

Demers-Payette, O., Lehoux, P. & Daudelin, G. (2016). Responsible research and innovation: a productive model for the future of medical innovation. *Journal of Responsible Innovation*, 3(3), 188–208. doi: 10.1080/23299460.2016.1256659.

Evers, H. D. (2002). *Knowledge Society and the Knowledge Gap*. New York: Zef Publishing.

Florén, H., Frishammar, J., Parida, V. & Wincent, J. (2017). Critical success factors in early new product development: a review and a conceptual model. *International Entrepreneurship and Management Journal*, 1–17.

Freeman, C. (1984). Prometheus unbound. *Futures*, 16(5), 494–507.

Genus, A. & Stirling, A. (2018). Collingridge and the dilemma of control: towards responsible and accountable innovation. *Research Policy*, 47(1), 61–69.

Grinbaum, A. (2013). The old-new meaning of researcher's responsibility. *Ethics & Politics*, XV(1), 236–250.

Gudowsky, N. and Peissl, W. (2016). Human centred science and technology: transdisciplinary foresight and co-creation as tools for active needs-based innovation governance. *European Journal of Futures Research*, 4(1), 8.

Gupta, A. K., Dey, A. R., Shinde, C., Mahanta, H., Patel, C. et al. (2016). Theory of open exclusive innovation for reciprocal, responsive and respect-ful outcomes: coping creatively with climatic and institutional risks. *Journal of Open Innovation: Technology, Market and Complexity*, 2(16), 1–15. doi: 10.1186/s40852-016-0038-8.

Hajer, M. A. (2009). *Authoritative Governance: Policy Making in the Age of Mediatization*. Oxford: Oxford University Press.

Inghelbrecht, L., Goeminne, G., van Huylenbroeck, G. & Dessein, J. (2016). When technology is more than instrumental: how ethical concerns in EU agriculture co-evolve with the development of GM crops. *Agriculture and Human Values*, 34(3), 543–557. doi: 10.1007/s10460-016-9742-z.

Irwin, A. (2006). The politics of talk: coming to terms with the 'new' scientific governance. *Social Studies of Science*, 36, 299–330.

Köhler, A. R. (2013). Material scarcity: a reason for responsibility in tech-nology development and product design. *Science and Engineering Ethics*, 19(3), 1165–1179. doi: 10.1007/s11948-012-9401-8.

Kohler-Koch, B. & Rittberger, B. (2006). The 'governance turn' in EU studies. *Journal of Common Market Studies*, Annual Review, pp. 27–49.

Kuzma, J. & Roberts, P. (2018). Cataloguing the barriers facing RRI in inno-vation pathways: a response to the dilemma of societal alignment. *Journal of Responsible Innovation*, 5(3), 338–346.

Le Feuvre, R. A., Carbonell, P., Currin, A., Dunstan, M., Fellows, D., Jervis, A. J., Rattray, N. J. W., Robinson, C. J., Swainston, N., Vinaixa, M., Williams, A., Yan, C., Barran, P., Breitling, R., Chen, G. G., Faulon, J.

L., Goble, C., Goodacre, R., Kell, D. B., Micklefield, J., Scrutton, N. S., Shapira, P., Takano, E. & Turner, N. J. (2016). Synbiochem: Synthetic Biology Research Centre, Manchester: a UK foundry for fine and speciality chemicals production. *Synthetic and Systems Biotechnology*, 1(4), 271.

Lubberink, R., Blok, V., van Ophem, J. & Omta, O. (2017). Lessons for responsible innovation in the business context: a systematic literature review of responsible, social and sustainable innovation practices. *Sustainability*, 9(5), 721.

MacCormack, A. & Verganti, R. (2003). Managing the sources of uncertainty: matching process and context in software development. *Journal of Product Innovation Management*, 20(3), 217–232.

Martin, R.B. (2010). The origins of the concept of 'foresight' in science and technology: an insider's perspective. *Technological Forecasting and Social Change*, 77, 1438–1447.

Martin, R. & Sunley, P. (2011). Conceptualizing cluster evolution: beyond the life cycle model? *Regional Studies*, 45(10), 1299–1318.

Meijboom, F. L. B., Visak, T. & Brom, F. W. A. (2006). From trust to trust-worthiness: why information is not enough in the food sector. *Journal of Agricultural and Environmental Ethics*, 19, 427–442.

Miller, G. & Wickson, F. (2015). Risk analysis of nanomaterials: exposing nanotechnology's naked emperor. *Review of Policy Research*, 32(4), 485–512.

Owen, R., Bessant, J. R. & Heintz, M. (2013)☐. *Responsible Innovation: Managing the Responsible Emergence of Science and Innovation in Society*. Chichester: John Wiley & Sons.

Owen, R., Stilgoe, J., Macnaghten, P., Gorman, M., Fisher, E. & Guston, D. (2013). A framework for responsible innovation. *Responsible Innovation: Managing the Responsible Emergence of Science and Innovation in Society*, 27–50. doi: 10.1002/9781118551424.ch2.

Pierce, R. L. (2013). Bridging current issues in science and society. *Biotechnology Journal*, 8, 875–877. doi: 10.1002/biot.201200264.

Rerimassie, V. (2016). Early engagement with synthetic biology in the Netherlands. Initiatives by the Rathenau Instituut. In Hagen, K., Engelhard, M. and Toepfer, G. (eds), *Ambivalences of Creating Life. Societal and Philosophical Dimensions of Synthetic Biology*. Berlin-Heidelberg: Springer Verlag, pp. 199–213.

Ribeiro, B., Bengtsson, L., Benneworth, P., Bührer, S., Castro-Martínez, E., Hansen, M. … & Shapira, P. (2018). Introducing the dilemma of societal alignment for inclusive and responsible research and innovation. *Journal of Responsible Innovation*, 5(3), 316–331.

Ries, E. (2011). *The Lean Startup: How Today's Entrepreneurs Use Continuous Innovation to Create Radically Successful Businesses*. London: Crown Books.

Rose, N. (2012). Democracy in the contemporary life sciences . *BioSocieties*, 7(4), 459–472.

Rothwell, R. (1994). Towards the fifth-generation innovation process. *International Marketing Review*, 11(1), 7–31.

Rowe, G. & Frewer, L. J. (2005). A typology of public engagement mechanisms. *Science, Technology, & Human Values*, 30(2), 251–290.

Ruggiu, D. (2015). Anchoring European governance: two versions of responsible research and innovation and EU fundamental rights as 'normative anchor points'. *NanoEthics*, 9, 217–235.

Schroeder, D., Dalton-Brown, S., Schrempf, B. & Kaplan, D. (2016). Responsible, inclusive innovation and the nano-divide. *Nanoethics*, 10, 177–188. doi: 10.1007/s11569-016-0265-2.

Selin, C., Kimbell, L., Ramirez, R. & Bhatti, Y. (2015). Scenarios and design: scoping the dialogue space. *Futures*, 74, 4–17. doi: 10.1016/j. futures.2015.06.002.

Silva, L., Bitencourt, C., Faccin, K. & Iakovleva, T. (forthcoming). The role of stakeholders in the context of responsible innovation to sustainable development goals: a meta-synthesis. *Sustainability*.

Spinello, R. A. (2003). The future of intellectual property. *Ethics and Information Technology*, 5, 1–16.

Stahl, B. C. & Coeckelbergh, M. (2016). Ethics of healthcare robotics: towards responsible research and innovation. *Robotics and Autonomous Systems*, 86, 152–161.

Stahl, B. C., Eden, G., Jirotka, M., Owen, R., Heintz, M. & Bessant, J. (2013). Responsible research and innovation in information and communication technology: identifying and engaging with the ethical implications of ICTs. In Owen, R., Bessant, J. and Heintz, M. (eds), *Responsible Innovation: Managing the Responsible Emergence of Science and Innovation in Society*. Chichester: John Wiley & Sons, pp. 199–218.

Stahl, B.C., O'Bach, M., Yahmaei, E., Ikonen, V., Chatfield, K. & Brem, A. (2017). The Responsible Research and Innovation (RRI) maturity model: linking theory and practice. *Sustainability*, 9(6), 1036–1055.

Stilgoe, J. (2012). Experiments in science policy: an autobiographical note. *Minerva*, 50, 197–204.

Stilgoe, J. (2016). Geoengineering as collective experimentation. *Science and Engineering Ethics*, 22(3), 851–869.

Stilgoe, J., Owen, R. & Macnaghten, P. (2013). Developing a framework for responsible innovation. *Research Policy*, 42(9), 1568–1580.

Taddeo, M. (2016). On the risks of relying on analogies to understand cyber conflicts. *Minds & Machines*, 26, 317–321. doi: 10.1007/s11023-016-9408-z.

te Kulve, H. & Rip, A. (2011). Constructing productive engagement: pre-engagement tools for emerging technologies. *Science and Engineering Ethics*, 17(4), 699–714.

Thapa, R., Iakovleva, T. & Foss, L. (under revisions). Responsible research and innovation: a systematic review of the literature and future research agenda, *European Planning Studies*.

Tidd. J. & Bessant, J. (2014). *Managing Innovation: Integrating Technological, Market and Organizational Change*, 5th edition. Chichester: John Wiley & Sons.

van Oost, E., Kuhlmann, S., Ordóñez-Matamoros, G.H. et al. (2016). Futures of science with and for society: towards transformative policy orientations. *Foresight – The Journal of Future Studies, Strategic Thinking and Policy*, 18(3), 276–296.

Vincent, B. B. (2013). Decentring nanoethics toward objects. *Etica e Politica*, 15(1), 310–320.

Voegtlin, C. & Scherer, A. G.(2015). Responsible innovation and the innovation of responsibility: governing sustainable development in a globalized world. *Journal of Business Ethics*, 143(2), 227–243 doi: 10.1007/s10551-015-2769-z.

Vogel, D. (2012). *The Politics of Precaution: Regulating Health, Safety, and Environmental Risks in Europe and the United States*. Princeton, NJ: Princeton University Press.

von Schomberg, R. (2011). *Towards Responsible Research and Innovation in the Information and Communication Technologies and Security Technologies Fields*. Luxembourg: Publications Office of the European Union.

von Schomberg, R. (2013). A vision of responsible innovation. In Owen, R., Bessant, J. and Heintz, M. (eds), *Responsible Innovation: Managing the Responsible Emergence of Science and Innovation in Society*. Chichester: John Wiley & Sons, pp. 51–74.

Walhout, B. & Konrad, K. (2015). Practicing responsible innovation in NanoNextNL. In Bowman, D., Dijkstra, A., Fautz, C., Guivant, J., Konrad, K., van Leute, H. and Woll, S. (eds), *Practices of Innovation, Governance and Action – Insights from Methods, Governance and Action*. Studies of New and Emerging Technologies 6. Berlin: IOS Press, pp. 53–68.

3. Challenges in healthcare – the changing role of patients

Elin M. Oftedal, Tatiana Iakovleva and John Bessant

3.1 HEALTHCARE SYSTEMS AND POWER BALANCE

As more people strive to live longer, healthier and more active lifestyles, healthcare concerns increase and so does the costs. Research reveals that healthcare costs and spending often rise at rates exceeding inflation, and they are expected to increase in the future. Keehan et al. (2017) estimate that aggregate healthcare spending in the United States will grow at an average annual rate of 5.8 per cent from 2015 through 2025, or 1.3 percentage points higher than the expected annual increase in the gross domestic product. The health sector faces many challenges today and represents a significant cost in the country's economy. Calculations show that average public spending on health services in most Western countries accounts for around 10 per cent of GDP. This situation is likely to worsen as a result of an ageing population, rising prices and increasing complexity associated with technology in the health services (Marmot et al., 2012), where there is pressure on delivering qualitatively good services to all, while significantly reducing costs.

Despite wide variations in healthcare funding systems across countries and regions, the underlying trends are the same. As pointed by Christensen, Waldeck and Fogg (2017), despite so many sophisticated medical technologies introduced every year, healthcare has not been disrupted to a significant degree. The reason being that technology has almost always been implemented in a sustaining manner in healthcare – primarily to help hospitals and doctors to solve the most complex problems. Thus, most technological enablers have failed to bring about lower costs, higher quality and greater accessibility (Christensen, Waldeck and Fogg, 2017). In this chapter we discuss the differences in healthcare systems in terms of concentration of power and the impact that might have on innovations. Further, we elaborate on the role of patients in the innovation process.

The cases in this book are situated in distinct healthcare contexts that might affect the nature of innovation and its diffusion in certain ways. Healthcare systems differ in terms of funding flows and power centres. The most important elements in a healthcare system are: (1) the government, (2) providers, (3) patients and (4) insurance companies. Traditionally, there are several types of healthcare models based on these four elements. The most basic system is when payment and services flows are directly between the providers and the patients/ users. This is called the 'out of pocket' model. Governments most often have a prominent position in healthcare systems to administer and manage the flow of finance and communication. Further, several systems include an insurance element, either taken up voluntarily or through obligation.

There are advantages and disadvantages with all the models. However, the patient has a weak position in all models. While different countries lean towards different models, few healthcare systems fit precisely within the parameters of a single model. Governments are concerned with the increasing portion of their economic resources that are spent on healthcare and therefore most national healthcare systems have experienced substantial reforms and ongoing improvements as governments try to balance resources between private and public income streams with the demand for health services and as such try to learn from good practices from other countries. Below we review the three most common models in which our empirical cases from this book are embedded. The illustration of the different models, are simplifications of very complex systems. We have done that in order to more easily compare the systems and to show specific qualities of each system. Therefore, none of the models will replicate the original system in a perfect manner.

3.2   THE BEVERIDGE MODEL

In the Beveridge System (see Figure 3.1), healthcare is provided and financed by the government through tax payments (CESifo, 2008). The power is centred at the government level, controlling the costs and the actors in the health sector. It is a top-down system, where government instructs the providers of healthcare through different mechanisms, and the providers instruct the patients. There is also an insurance system, but it is an addition to the system, not central, and is used in varying degrees. Several countries use this system, including Great Britain, Spain, most of Scandinavia and New Zealand, Hong Kong and Cuba. Here, hospitals and clinics are mainly owned by the government; and the role of private actors varies. Seen from a perspective of the larger society, the benefits of using this model is that costs per capita are low since the government, as the sole payer, controls what doctors can do

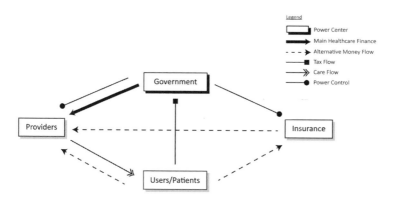

Figure 3.1   The Beveridge Model

and what they can charge. This model is also associated with high efficiency scores (Cylus, Papanicolas and Smit, 2016; OECD/WHO/World Bank Group, 2018). Nevertheless, there are also challenges associated with this system: (1) It is demanding for the government to control cost and (2) There is a lack of competition, which can lead to a lack of effectiveness. In terms of healthcare, the demographic changes are becoming an increasing burden on tax revenues both quantitatively (more older and illness-prone people) and qualitatively (more expensive medical services and technology). The problem here is the financing of the healthcare system must compete for tax allocation with other policy areas.

To meet these challenges it is necessary to have good bureaucratic routines. From a patient's point of view, the benefits are a feeling of safety among citizens as patients do not have to worry about financing their treatment. On the negative side, patients have little opportunity to influence as the power balance is uneven and there is a lack of the usual market mechanisms. Patients cannot freely choose their caregiver or have any influence on treatment or tools that can facilitate their everyday situation, but must rely on the caregiver assigned to them by the system.

3.3   THE BISMARCK MODEL

The Bismarck Model (see Figure 3.2) is a multi-payer healthcare system paid for by a combination of statutory health insurance officially called 'sickness funds' and private health insurance colloquially also called '(private) sickness funds' (CESifo, 2008). Bismarck-type statutory health insurance plans have

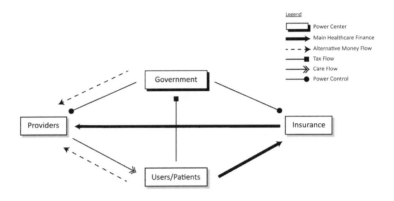

Figure 3.2   The Bismarck Model

to cover everybody, and do not make a profit. The financing is via contributions, graduated according to income. The contributions to be paid are based
on wages or salaries. High-income earners and those who demand better
healthcare than the standard service can acquire more expensive, private health
plans. Generally, these profit-driven, private insurers offer better, more extensive coverage than the statutory health insurance. Another interesting point is
that even if this is a universal healthcare system, doctors and hospitals tend to
be private in Bismarck countries.

Nevertheless, the balance of power is with the government which has tight
control on prices (CESifo, 2008). The tight regulation gives government
much of the cost-control that the single-payer Beveridge Model provides. The
Bismarck Model is widely used, applied in countries such as Germany, France,
Belgium, the Netherlands, Japan, Switzerland, and, to a degree, in Latin
America. The Bismarck model has some clear advantages but also some limitations. The advantages are that: (1) every citizen is insured, giving citizens
the same protection as in the Beveridge System; (2) since it is an insurance
system, the patient has a greater level of choice than in the Beveridge system;
and (3) the system also has high efficiency results. Disadvantages are: (1) it is
a fairly expensive system, both for patient and for government; and (2) social
insurance contributors must support an increasing number of people who no
longer contribute. This is something the Bismarck System did not envision
(CESifo, 2008).

3.4   THE PRIVATE INSURANCE MODEL

This model (see Figure 3.3) is characterized by employment-based or individual purchase of private health insurance financed by individual and employer contributions. The main power in this system is held by the insurance companies and the providers. Service delivery and financing are owned and managed by the private entities operating in an open market economy. The pros are that there is: (1) larger choice for the patient, as they are free to choose doctor and medical offices; and (2) high competition leads to innovation and state-of-the-art treatment. The drawbacks to this system are: (1) high cost and complex systems; (2) expensive for patients, as it calls for both higher out of pocket expense in addition to a high insurance fee; (3) lack of universal coverage, leading to lack of protection for the entire population; and (4) insecurity even for insured citizens, since they do not want use their insurance for lesser cases. This can become very expensive for both the patient and society as citizens rely on self-care instead of insurance. They often wait too long to seek help and then there is a larger emergency operation that will be more costly (Schoen et al., 2018). Moreover, as insurance is on a case by case basis, a challenge for most new healthcare innovations is the limited reimbursement for the treatment. As a result, even if the solution exists it is rarely an option for the majority of the people.

While different countries lean towards different versions of each system, all of them have challenges to bring affordable care of high quality to all citizens at a reasonable costs. While the government-centric Beveridge Model provides available care to all citizens its leaves little choices for differentiated

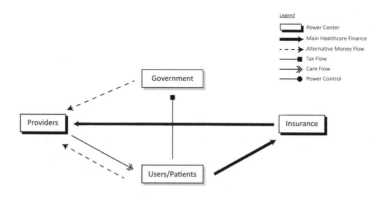

Figure 3.3   The Private Insurance Model

care when needed. Furthermore, as the demographic changes, this model is in some instances becoming very expensive to finance. The Bismarck Model provides more choices to citizens and at the same time ensures everyone is covered, but it is costly and insurance agencies act as third parties that might dictate the rules of the game for patients. The Private Insurance Model offers ample choices to patients, but at such high costs that services may become unavailable for some citizen groups. The patient has a low degree of power in all of these systems. The power centre in the Beveridge or Bismark models lie in government institutions, while in the Private Insurance Model power it is concentrated among providers and insurance companies. This leads to the ignorance of user voices and transformation of systems to satisfy the needs of 'power centres'. In the Private Insurance Model patients can, to certain degree, influence decisions through signalling their preferences, creating an incentive for competition. In theory, this system would be conducive to innovation. However, in reality, the providers and the insurers drive incremental change that does not disrupt the market and which could make healthcare affordable and available for everyone in need.

3.5   THE PATIENT ROLE IN HEALTHCARE FROM AN INNOVATION PERSPECTIVE

The healthcare industry has long relied on traditional and linear models of innovation; basic and applied research followed by development and commercialization (Omachonu and Einspruch, 2010). This is a technology-push model of innovation that has the aim of sustaining the current market players in the industry (Christensen, Waldeck and Fogg, 2017). Healthcare is all about the patient – yet for so much of its evolution the patient has been treated as a passive recipient. Traditionally, patients have been a passive recipient of healthcare, and have been a victim of the circumstances rather than a powerful actor, leading to one-way communication relationships between healthcare professionals and patients (Epstein et al., 1993). Berry et al. (2017) discuss how patients are often reluctant to assert their interests in the presence of clinicians, whom they see as experts. As a result, many patients are prone to the alleged 'hostage bargaining syndrome' (HBS), meaning patients behave as if negotiating for their health from a position of fear and confusion. It may manifest as understating a concern, asking for less than what is desired or needed, or even remaining silent against one's better judgment. When HBS persists and escalates, a patient may succumb to learned helplessness, making his or her authentic involvement in shared decision making almost impossible (Berry et al., 2017). To challenge the HBS syndrome, clinicians must aim to be sensitive to the power imbalance inherent in the clinician–patient relationship: clinicians often have the power to reverse this by appreciating, paradoxically,

Table 3.1   *Users through innovation lenses*

Type of patient	Description	Questions to be addressed
The Informed Patient	User is actively using sources of information to understand own circumstances	How to prevent 'disinformed' patient
The Involved Patient	Patient is actively involved by the healthcare sector	How to create mechanisms to involve and listen to users' perspective? How to most efficiently reflect on user feedback?
The Innovative Patient	Patient is actively innovating to find new solution for the problems	How to create space for innovation created by the patient innovators/entrepreneurs in this sector?

how patients' perceptions of their power as experts play a central role in the care they provide. Today, many adults are understandably wary of physicians' advice. Long-standing relationships between patients and doctors are scarce, if not unusual. Trust comes in knowing a person well and over time, an aspect of medical care that's lost when insurance plans change and is further damaged by a breed of younger physicians who, with a keen instinct for self-preservation, may sensibly choose to limit the hours they work rather than quit the field entirely (Birkhäuer et al., 2017)

However, new digital technologies open up opportunities to change this situation. Digitalization of healthcare empowers patients to shape and direct the technologies in their own interests (Bos et al., 2008). The new technologies give rise to the idea that healthcare should be all about the patient. New technology enables patients to gain a more active role in healthcare systems. There is an increasingly vast volume of health information available to patients and doctors, making it easier than earlier for patients to develop their own diagnosis (Giustini, 2005). This information may enable patients to be well informed. An example of this is found in the Norwegian television program 'What's wrong with you?' where a team of 'laymen' competes with a group of select doctors about diagnosing a patient. The laymen can use search engines such as Google, while the doctors can only use their knowledge as physicians. The score between the groups is often even, although the doctors usually had a slight advantage. This example, however, proves the democratization of knowledge and a new information flow.

The rise of digital technologies in the healthcare sector leads not only to technological development, but also to a change in the state-of-mind, a way of thinking, an attitude towards healthcare and role of different stakeholders in it. Today one can observe a greater commitment for networked, global thinking, to improve healthcare locally, regionally, and worldwide by using information and communication technology (Eysenbach and Jadad, 2001). This opens up opportunities for increasing participation of consumers in digital health outside of the hospital setting. Modern society is characterized by its increasing acceptance and use of digital technology, exemplified by the preference to receive information through digital routes. The ease of information sharing and availability of information challenge the established view on patients in the healthcare sector as passive receivers of services. Patients today are not only able to search for information with regard to their situation, they also become active discussants of their situation with healthcare professionals, and at some extremes they can become providers and a source of innovative solutions. Thus, opening up the user dimension offers an important contribution, moving the debate from considering patients as a homogenous group to viewing them as distributed across a spectrum:

1. The 'informed patient', equipped to use technology based on improved understanding;
2. The 'involved patient', playing an active role within a wider healthcare delivery system and enabled to do so by technology;
3. The 'innovating patient', providing ideas of their own based on their deep understanding of their healthcare issue.

Below we broadly discuss this three core groups of patients viewed through the lens of innovation (see Table 3.1).

3.5.1   The Informed Patient

Communication in medicine is work in progress. We are immersed in data, directly delivered by the Internet into our homes. We can access reams of raw, un-spun research results; these turn up in government-funded, transparency-directed databases and in the online supplements to serious scientific journals (Schattner, 2010). At the other extreme, we can take in processed, neat-and-tidy bits of medical knowledge from commercials and TV documentaries. The Internet allows us to search and get answers to questions in a matter of seconds. This is of course true for the patients facing the questions on how to treat their illness. We can speak about the emerging concept of an informed patient. This means a patient who's knowledgeable about his or her condition and engaged in health decisions. This can ease the job for the doctor and bring more knowledge into the health system. The informed patient today has much better knowledge access in comparison to even a decade ago, with general search as well as specialized databases available to service an information search. Better understanding of symptoms and treatments, availability of feedback through fora and social media contribute to balancing the power in the health professional–patient relationship. As argued by recent research, there is new potential for information parity and equality and, as a result, emerging authority for patients to be the decision maker (Topol, 2015). The informed patient might challenge the power of the relationship through questions, obtain a better understanding about treatments, consequences and choices available.

In fact, health literacy is increasingly recognized as essential for successful access, navigation, self-care and management of health and wellness (Manganello and Shone, 2013). The challenge is that there is a large gap between the literacy demands of the health system and the health literacy skills of most people. Low health literacy is associated with less knowledge of diseases and self-care, worse self-management skills, lower medication compliance rates, higher rates of hospitalization and poorer health outcomes. Health literacy problems are magnified as patients are increasingly asked to take more responsibility for their health in a healthcare system that is increasingly complex, specialized and technologically sophisticated (Manganello and Shone, 2013).

In our book, we illustrate the efforts to facilitate dialogue between patients and physicians. For example Chapter 6 describes the University of Virginia system called MyChart, which allows patients to stay informed and in touch with physicians. This is one of many examples of how digitalization enables the process of informing and educating the patient. In Chapter 14 examples are provided of digital applications that allow users to become truly informed and even involved patients. These cases suggest a new dynamic in healthcare – an open, modern discussion between patients and physicians. Professionalism is the basis of medicine's contract with society, and the principle of patient autonomy is a fundamental value of the medical profession. This allows for a stronger patient autonomy, empowering patients to make informed decisions about their health.

3.5.2 The Involved Patient

The development of a patient-centred approach to medicine is gradually allowing more patients to be involved in their own medical decisions. According to the predominant culture, research is performed on patients, not with patients (Thornton, 2014). Thus, patients continue to be regarded as a source of data and not as the true protagonists in the process. However, patient involvement is crucial for identifying the questions to ask and the outcomes to assess (Sacristán et al., 2016). Digitalization not only provides users with information, it also empowers them to share their feedback in a timely manner. A wide variety of feedback emphasized by digital technologies can be observed – from tele media communications with doctors and nurses to forum discussions on social media. This allowed utilization of patients' feedback to the mutual benefit of healthcare provider and patient. This also made it possible to improve and adapt products, services or processes within healthcare settings (Drejer, 2004). Recent studies demonstrate that patients' engagement contributes to increased innovation in medical units (Hibbard and Greene, 2013). From being an informed patient, users now become co-creators in the innovation process through engagement. However, user feedback and ideas have to be heard and transformed into outcomes. Thus, this co-creation process is dependent not only upon patients themselves, but also on the willingness of the organization (for example a hospital) to absorb and react on this type of feedback. This requires responsiveness and reflectiveness to be in place to ensure the responsible innovation process (Stilgoe, Owen and Macnaghten, 2013; Bessant et al., 2017).

Today it is increasingly common to involve patients or patient advocacy groups in study design. Patient advocacy groups now claim that their opinions must have greater influence on the decisions that affect them, which is reflected in the phrase 'nothing about me without me'. The development of new healthcare management models where patients become clients and the enormous expansion of information technology are additional factors that contribute to accelerate this change (Porter, 2010). Patient-centred medicine cannot be practised without patients participating in their own healthcare decisions and in the research that informs such decisions (Sacristán et al., 2016).

Although this cultural shift is beginning to change the way we understand healthcare, it is not having the same impact on the research process (Lloyd and White, 2011). This may be because society does not see patient responsibility to participate in research as obvious as the responsibility to participate in their own medical care (Tinetti and Basch, 2013).

Vahdat et al. (2014) found that an effective relationship of a healthcare provider with patients is an important contributing factor of patient involvement in decision making. In addition, a literature review confirms that patients, in their journey through the healthcare system have the right to be treated respectfully and honestly, and where possible, be involved in their own healthcare decisions. For patients' participation, mutual communication between treatment team and the patient is necessary, so that information and knowledge could be shared between them, giving the patient a sense of control and responsibility, and thus involving the patient in care activities (mental or physical), to benefit and rehabilitate from this involvement (Fleurence et al., 2014; Basch, 2013).

Several chapters of our book illustrate the situation of the involved patient. Chapter 5 demonstrates that patient feedback in many cases can stimulate organizational improvements and service innovations in hospitals. Using the particular example of a detailed longitudinal case of the development and diffusion (with subsequent modification and 're-innovation') of a digital health information platform, Chapter 7 identifies a number of key points at which the innovation concept 'pivoted' to reflect new information, some of which resulted from a wider level of inclusion. Co-creation of digital innovative solutions for elderly citizens by different stakeholder groups, including relatives, volunteers and healthcare professionals is described in Chapter 11.

3.5.3 The Innovative Patient

An alternative emerging at healthcare institutions worldwide is human-centred design and co-creation, a set of approaches that can accelerate and humanize healthcare innovation. This model is not just about getting greater patient feedback during the innovation process. Patients are co-designers, co-developers, and increasingly more responsible for their own and collective health outcomes. Who knows the challenges of living with diabetes or cancer better than those who experience it? Patients with chronic diseases are often forced to find solutions for their conditions, if solutions for their problems are not available already on the market. There is growing evidence that many patients and caregivers develop solutions to cope with their health disorders (Canhão, Zejnilovic and Oliveira, 2017; Zejnilovic et al., 2016). Canhão, Zejnilovic and Oliveira (2017) report on studies of patient- and caregiver-originated innovations. In their research they found a variety of ingenious solutions to daily problems, previously unknown therapies and treatments, and even new ideas for medical devices. Surprisingly 8 per cent of the total rare disease patients and caregivers had developed innovative solutions that even medical experts evaluated as novel. They concluded that patients and caregivers around the globe may represent a tremendous source of knowledge on how to improve care.

Chapter 4 of this book describes Oliveira and Canhão's ongoing research. They have collected 650 innovations developed by patients and caregivers, and have posted them on the Patient Innovation website. Some of them are technically very simple, but offer great value to patients and their families, while others are more advanced and complex. The vast majority of them focus on increasing patients' autonomy and quality of life. Evidence also shows that these innovations have low diffusion rates. This is, especially, a situation in the segment of rare diseases and chronic needs – niche markets, unattractive to stakeholders in the healthcare industry – where patients are pioneers regarding innovation when compared to commercial producers, which can help close gaps in the delivery of medical care (Habicht, Oliveira and Shcherbatiuk, 2012). In sharp contrast to the frequency of patient-to-patient sharing (88 per cent of those who shared solutions), only 6 per cent of patients reported describing their innovations to their clinicians, where reduced appointment time or lack of confidence are possible reasons (Oliveira et al., 2015). In other words, the development and diffusion of patient-developed solutions are confined to small groups of peers, and these activities are hidden from the healthcare system and the general population. Canhão, Zejnilovic and Oliveira (2017) claim that the absence of diffusion signify barriers to sharing as the inventors may lack time, skills, and opportunities to embark on the long process between idea and successful commercialization when the profit potential is limited or the inventors' motivations lie elsewhere. Simpler solutions may not be shared because the innovators do not have contact with a wider community that would benefit. While some trade their innovations in patient support communities, others are less outgoing. They conclude that since all patients interact with the medical community, the latter can be essential in the distribution of new ideas. Physicians and other members of the care team could be attentive to innovations coming from their patients, and even active in reaching new patient groups which the innovation may benefit. Patients are not the only innovators, of course. Many people, across all fields, innovate to address their unmet needs, in a model that MIT researcher Eric von Hippel calls "user innovation" (von Hippel, DeMonaco and de Jong, 2017). Traditional "producer innovation" co-exists, and sometimes competes, with innovations that spring out of users and communities. However, while producer innovations have a well-worn path for distribution through standard commercial channels, consumer-originated innovations are substantially less likely ever to leave their inventors' hands, even when they address common or urgent needs.

3.6   THE FUTURE OF DIGITAL HEALTH

As was demonstrated in this chapter, healthcare systems have power centres where power resides with government, providers or insurance companies. Patients, while being a core concern of healthcare systems, are not in the position to directly affect implementation of much-needed innovative solutions in this sector. It seems that despite the recognition of the need for patient empowerment, innovations in healthcare are developed along a sustaining trajectory, focusing primarily on satisfying providers' needs.

On the other hand, today one can observe a move towards patient-centred medicine that aims to provide the best healthcare for each individual patient, taking his or her goals, preferences, and values into account. Doctor–patient relationships are changing, and concepts such as shared decision making and patient empowerment are becoming a reality. Digital technologies contribute to this radical change, they help to break established structures and practices. They also contribute to new approaches in healthcare and shifting the power balance in favour of patients. The pace of these 'disturbances' continues to accelerate and it causes extensive changes across different parts of the businesses (Bessant et al., 2017). Today we also observe the surge of entrepreneurial activity around digital health (Lupton, 2014). Many new entrants are competing with mainstream players in healthcare in exploring how new technological promise might be harnessed and a new dynamic injected into an old market (Brynjolfsson and McAfee, 2014). For instance, digital technologies enable the management and processing of medical data previously locked away in handwritten records at much faster speeds (Steinhubl and Topol, 2015). Furthermore, it is now possible to digitize key functions and capabilities of physical products with remarkable price/performance ratios disrupting the entire business model (Yoo et al., 2012).

Digitization increases the possibility of participating in digital health for all stakeholders. Acceptance of digital solutions among health professionals and healthcare can open many exciting opportunities – to deliver higher quality healthcare for more people and for less cost. At the same time, radical innovations cause resistance to change in existing systems. As we demonstrated earlier in this chapter, existing systems do not always consider patients as power actors; they are also complex systems with a low rate of change.

Institutional theory argues that any structures are embedded into the context (Scott, 2014). Such context consists of regulatory, cognitive and normative dimensions that affect the behaviour of organizations and individuals. Regulatory dimensions relate to formal rules and laws, which are quite restrictive in healthcare and take a long time to change. For example, there are significant barriers in the form of regulations and privacy, security and the need for confidentiality. Thus, there is little room for innovators to contribute with new products and services and even less opportunity for patients who lack experience in business activities to contribute to this process. It is also clear that the entrepreneurial path is exceptional for patient innovators. Statistics shows that only 5 to 15 per cent of any given country dare to take an entrepreneurial path (GEM, 2017). Even though we know that innovative patients constitute considerable numbers, it is reasonable to think that taking actions to commercialize innovations might not be a priority for chronically sick individuals (Oliveira et al., 2018). We need rules and regulations that are favourable towards disruptive innovations in digital health, which create spaces for entrepreneurs and patient innovators to bring their solutions to the market, and that allow user voices to be heard and reflected upon.

A second context dimension according to institutional theory is the cognitive dimension. It stands for knowledge and understanding of the new role patients have in healthcare due to the digitization processes. There is a big difference between countries in terms of digital skills, but even in countries where the population generally has good access to digital media, one can talk about a 'digital divide'. Technology is changing so quickly that some populations (for example, elderly people with mental disorders, deaf patients) cannot keep up with developments and consequently cannot benefit from new digital solutions. However, the knowledge of opportunities offered by new digital tools in health and welfare services is important both for patients and health professionals. The more knowledge about new digital solutions healthcare professionals have, the easier it is to accept and apply innovations.

The final dimension in relation to institutional theory is normative. It includes values and accepted norms of behaviour among the population as well as health professionals. It emerges as a significant barrier to patient-developed innovations and also generally for new digital solutions introduced on the market. The patient is often seen as a passive recipient of assistance, assuming that the knowledge lies with health professionals and thus it is difficult to break this barrier, both for patients when they are afraid to receive negative feedback and for health professionals as they are not used to grasping new input and knowledge from patients, as discussed in Chapter 5 in this book. General barriers associated with digital solutions, regardless of whether they are developed by patients or not, are a lack of trust by both healthcare professionals and users. Recent research shows that there are four different barriers identified in implementing well-being technology at nursing homes in Norway: organizational resistance, cultural resistance, technological resistance and ethical resistance (Nilsen et al., 2016). It is pointed out that employees in nursing homes perceived more threats in implementing digital technology – threats to stability and predictability, threats to role and group identity, threats to basic health values.

For the time being, digital solutions do not necessarily require medical tests, such as those required for commercialization of medication. This leads to low confidence among doctors and health professionals for digital health products, as pointed in Chapter 14 of this book. Digital aids within health and welfare services often require changes in behaviour, and there are many personal factors that can affect the outcome. Thus, it may be harder to achieve the desired result from treatment carried out with the help of such digital solutions.

3.7   CONCLUSION

In this chapter we have considered the need for change of the healthcare systems and recognition of patients' contributions to foster innovations in the healthcare sector. The majority of patients can be labelled as informed and involved users, who actively use digital channels to obtain information that can help them in their situation. They have both the skills and knowledge they want to share with other patients and not least with healthcare professionals. Given that patient participation results in improved health outcomes, enhanced quality of life, more compliance and cost effectiveness of services, patients should be regarded as equal partners in healthcare. They will actively participate in their own healthcare process, and would more carefully follow their own care.

However regulative, cognitive and normative constraints may prevent existing systems from fully integrating this potential. The risk with powerful new technologies that are pushed to the market in a traditional way is that patients might become further marginalized – healthcare is something done to them and delivered in an increasingly non-human fashion. An alternative view is to put the patient at the centre and explore models that would empower them to shape and direct the technologies in their own interests.

Therefore the concept of responsible innovation in healthcare becomes timely. As stakeholder inclusion constitutes the core of the responsibility concept, empowering the patient should truly become a driver of innovation in healthcare. This participation has the objective of broadening visions, purposes, issues and dilemmas for wide and collective deliberation through the processes of dialogue, engagement and debate (Owen et al., 2013). Participation of patients can improve consultation, seeking opinions, or use of their actual and potential abilities. Further, participation might result in better rehabilitation of patients. In this case, by acquiring knowledge, skills, and self-confidence, patients will be able to take care of their own health and manage to live life competently (Vahdat et al., 2014).

Seeking to transform the world, the United Nations developed 17 Sustainable Development Goals (SDG), that offer a plan of action for people, planet and prosperity (United Nations, 2015). Those goals are developed to, until 2035, stimulate action in critical areas, including availability of affordable healthcare services. The stakeholder participation is heavily connected to SDG17, highlighting the search for the multi-stakeholder partnerships to "mobilize and share knowledge, expertise, technology and financial resource" (United Nations, 2015, p. 27), and at the same time "encourage and promote effective public, public–private and civil society partnerships, building on the experience and resourcing strategies of partnerships". Responsible innovation evokes a collective duty of care: a commitment to rethink the purposes and impacts of innovation, as well as a reflection on how to make its pathways sensitive to uncertainty (Mejlgaard and Bloch, 2012).

Thus, planning and providing patient-oriented healthcare based on the opinions, needs, and preferences of patients are recommended. Politicians, policymakers, healthcare professionals and broader society must challenge their view on existing healthcare systems and ask critical questions such as: how to enable users to become more involved in innovations; and about how society can exploit the great potential of patients and users. Countries must find alternative methods to combat the rising costs of care. They must do the research to find funding, grants and contributors to help them conduct research, set up programmes and implement processes at the pace of change. Digital technologies can be a helping hand in this process, but their implementation should be done in a responsible way.

REFERENCES

Basch, E. (2013). Toward patient-centered drug development in oncology. *New England Journal of Medicine*, 369(5), 397–400.

Berry, L. L., Danaher, T. S., Beckham, D., Awdish, R. L. & Mate, K. S. (2017). When patients and their families feel like hostages to health care. *Mayo Clinic Proceedings*, 92(9), 1373–1381.

Bessant, J., Alexander, A., Wynne, D. & Trifilova, A. (2017). Responsible innovation in healthcare: the case of health information TV. *International Journal of Innovation Management*, 21(8), 1740012.

Birkhäuer, J., Gaab, J., Kossowsky, J., Hasler, S., Krummenacher, P., Werner, C. & Gerger, H. (2017). Trust in the health care professional and health outcome: a meta-analysis. *PloS One*, 12(2), e0170988. doi: 10.1371/journal.pone.0170988.

Bos, L., Marsh, A. Carroll, D., Gupta, S. & Reeds, M. (2008). Patient 2.0 Empowerment. Conference: Proceedings of the 2008 International Conference on Semantic Web & Web Services, SWWS 2008.

Brynjolfsson, E. & McAfee, A. (2014). *The Second Machine Age: Work, Progress, and Prosperity in a Time of Brilliant Technologies*. London: WW Norton & Company.

Canhão, H., Zejnilovic, L. & Oliveira, P. (2017). Revolutionising healthcare by empowering patients to innovate. *European Medical Journal – Innovations*, 1(1), 31–34.

CESifo (2008). DICE Report 4/2008 (Winter), Ifo Institute for Economic Research, Munich, 2008, 01–75.

Christensen, C., Waldeck, A. & Fogg, R. (2017). How disruptive innovation can finally revolutionize healthcare. Innosight executive briefing.

Cylus, J. Papanicolas, I. & Smit, P. C. (2016). Health system efficiency: how to make measurement matter for policy and management. *Health Policy Series*. Retrieved 13 March 2019 from http://www.euro.who.int/__data/assets/pdf_file/0004/324283/Health-System-Efficiency-How-make-measurement-matter-policy-management.pdf.

Drejer, I. (2004). Identifying innovation in surveys of services: a Schumpeterian perspective. *Research Policy*, 33(3), 551–562.

Epstein, R. M., Campbell, T. L., Cohen-Cole, S. A., McWhinney, I. R. & Smilkstein, G. (1993). Perspectives on patient–doctor communication. *Journal of Family Practice*, 37, 377–377.

Eysenbach, G. and Jadad, A. R. (2001). Evidence-based patient choice and consumer health informatics in the Internet age. *Journal of Medical Internet Research*, 3(2), e19.

Fleurence, R. L., Forsythe, L. P., Lauer, M., Rotter, J., Ioannidis, J. P., Beal, A. ... & Selby, J. V. (2014). Engaging patients and stakeholders in research proposal review: the patient-centered outcomes research institute. *Annals of Internal Medicine*, 161(2), 122–130.

GEM (2017). Retrieved 30 November 2018 from https://www.gemconsortium .org/.

Giustini, D. (2005). How Google is changing medicine. *British Medical Journal*, 331, 1487.

Habicht, H., Oliveira, P. & Shcherbatiuk, V. (2012). User innovators: when patients set out to help themselves and end up helping many. *Die Unternehmung*, 66(3), 277–294.

Hibbard, J. H. & Greene, J. (2013). What the evidence shows about patient activation: better health outcomes and care experiences; fewer data on costs. *Health Affairs* (Millwood), 32(2), 207–14. doi: 10.1377/hlthaff.2012.1061.

Keehan, S. P., Stone, D. A., Poisal, J. A., Cuckler, G. A., Sisko, A. M., Smith, S. D. ... & Lizonitz, J. M. (2017). National health expenditure projections, 2016–25: price increases, aging push sector to 20 percent of economy. *Health Affairs*, 36(3), 553–563.

Lloyd, K. and White, J. (2011). Democratizing clinical research. *Nature*, N474, 277–278.

Lupton, D. (2014). Beyond techno-utopia: critical approaches to digital health technologies. *Societies*, 4(4), 706–711. https://doi.org/10.3390/soc4040706.

Manganello, J. A. & Shone, L. P. (2013). Health literacy: research FACTS and findings. ACT for Youth Center of Excellence. Ithaca, NY. Retrieved 13 March 2019 from http://www.actforyouth.net/resources/rf/rf_health -literacy_0513.pdf.

Marmot, M., Allen, J., Bell, R. and Goldblatt, P. (2012). Building of the global movement for health equity: from Santiago to Rio and beyond. *Lancet*, 379, 181–188. doi: 10.1016/S0140-6736(11)61506-7.

Mejlgaard, N. & Bloch, C. (2012). Science in society in Europe. *Science and Public Policy*, 39(6), 695–700. doi: 10.1093/scipol/scs087.

Nilsen, E. R., Dugstad, J., Eide, H., Gullslett, M. K. & Eide, T. (2016). Exploring resistance to implementation of welfare technology in munic-ipal healthcare services: a longitudinal case study. *BMC Health Services Research*, 16(1), 657.

OECD/WHO/World Bank Group (2018). *Delivering Quality Health Services: A Global Imperative*, WHO, Geneva 27. https://doi.org/10.1787/ 9789264300309-en.

Oliveira, P., Zejnilovic, H. & Canhão, H. (2017). Challenges and opportunities in developing and sharing solutions by patients and caregivers: the story of a knowledge commons for the patient innovation project. *Governing Medical Knowledge Commons*, 301.

Oliveira, P., Zejnilovic, L., Azevedo, S., Rodrigues, A,M. & Canhão, H. (2018). Peer-adoption and development of health innovations by patients: a national representative study of 6204 citizens, *JMIR Preprints*, 30/07/2018:11726. doi: 10.2196/preprints.11726, URL: http://preprints.jmir.org/preprint/11726.

Oliveira, P., Zejnilovic, L., Canhão, H. & von Hippel, E. (2015). Patient innovation under rare diseases and chronic condition. *Orphanet Journal of Rare Diseases*, 9(10), 41.

Omachonu, V. K. & Einspruch, N. G. (2010). Innovation in healthcare delivery systems: a conceptual framework. *The Innovation Journal: The Public Sector Innovation Journal*, 15(1), 1–20.

Owen, R., Stilgoe, J., Macnaghten, P., Gorman, M., Fisher, E. & Guston, D. (2013). A framework for responsible innovation. *Responsible Innovation: Managing the Responsible Emergence of Science and Innovation in Society*, 27–50. doi: 10.1002/9781118551424.ch2.

Porter M. (2010). What is value in health care? New England Journal of Medicine, N363, 2477–2481.

Sacristán, J. A., Aguarón, A., Avendaño-Solá, C., Garrido, P., Carrión, J., Gutiérrez, A. ... & Flores, A. (2016). Patient involvement in clinical research: why, when, and how. *Patient Preference and Adherence*, 10, 631.

Schattner, E. (2010) Are you an informed patient? Blog. Retrieved 3 November 2018 from https://www.huffpost.com/entry/are-you-an-informed-patie_b_404611.

Schoen, C., Davis, K., Buttorf, C. and Willink, A. (2018). Medicare benefit redesign: enhancing affordability for beneficiaries while promoting choice and competition. *Issue Brief (Commonwealth Fund)*, 2018, 1.

Scott, W. R. (2014). *Institutions and Organizations: Ideas, Interests and Identity*. London: Saga Publishing.

Steinhubl, S. R. & Topol, E. J. (2015). Moving from digitalization to digitization in cardiovascular care: why is it important, and what could it mean for patients and providers? *Journal of the American College of Cardiology*, 66(13), 1489–1496.

Stilgoe, J., Owen, R. & Macnaghten, P. (2013). Developing a framework for responsible innovation. *Research Policy*, 42(9), 1568–1580.

Thornton, S. (2014). Beyond rhetoric: we need a strategy for patient involvement in the health service. *British Medical Journal*, N348, g4072.

Tinetti, M. & Basch, E.(2013). Patients' responsibility to participate in decision making and research. *Journal of the American Medical Association*, N309, 2331–2332.

Topol, E. (2015). *The Patient Will See You Now: The Future of Medicine is in Your Hands*. New York: Basic Books.

United Nations. (2015). Transforming our world: the 2030 agenda for sustainable development. Resolution adopted by the General Assembly. Retrieved 30 November 2018 from https://sustainabledevelopment.un.org/post2015/transformingourworld/publication.

Vahdat, S., Hamzehgardeshi, L., Hessam, S. & Hamzehgardeshi, Z. (2014). Patient involvement in health care decision making: a review. *Iranian Red Crescent Medical Journal*, 16(1), e12454.

von Hippel, E., DeMonaco, H. & de Jong, J. P. (2017). Market failure in the diffusion of clinician-developed innovations: the case of off-label drug discoveries. *Science and Public Policy*, 44(1), 121–131.

Yoo, Y., Boland Jr., R. J., Lyytinen, K. & Majchrzak, A. (2012). Organizing for innovation in the digitized world. *Organization Science*, 23(5), 1398–1408.

Zejnilovic L., Oliveira, P. & Canhão, H. (2016). Innovations by and for the patients: and how can we integrate them into the future health care system. In H. Albach, H. Meffert, S, Pinkwort, R. Reichwald and W. von Eiff (eds), *Boundaryless Hospital*. Berlin: Springer-Verlag, 341–357.

4. Empowering patients to innovate: the case of Patient Innovation

Pedro Oliveira, Salomé Azevedo and Helena Canhão

4.1   INTRODUCTION

Placing the patient at the centre of healthcare management, empowering his preferences and decisions, valuing his role in follow-up and disease management is a recent trend. Patients and patient associations are leading awareness campaigns, education and training and even trials with new drugs and strategies. Although, innovation has until recently been seen as a product and output predominantly from labs and companies, recent research has shown that many patients and non-professional caregivers develop solutions to cope with their health disorders (Oliveira et al., 2015). However, patient innovation has low diffusion beyond its developer and consequently offers limited benefit to other potential users.

In this chapter we use a two-stage approach to discuss three research questions: Do patients, caregivers and collaborators play a role in health innovation? What are their motivations to innovate and share? Could an online open platform promote diffusion of patient-driven health innovations?

In the first part of the chapter we discuss these questions by using 11 case studies focusing on innovations by patients and caregivers. In the second part of the chapter we analyse data from the open, free, online platform www.patient-innovation.com, for patients and caregivers to share their innovations.

The case studies provide a variety of examples of individuals developing solutions to health-related problems that are not addressed by current medical practice. The solutions have different levels of complexity and respond to distinct diseases or activities. Motivations to innovate vary among subjects and diffusion occurs through a variety of channels. Some patient innovators become entrepreneurs and commercialize their solutions. In order to understand the degree of diffusion of these innovations we extract and analyse data from the free, open, online platform Patient Innovation. Currently this platform hosts more than 850 innovations developed by patients and caregivers worldwide and provides measures of its usage and impact. We analysed 734 solutions screened, cleared and published between February 2014 and December 2017.

The concept of patient innovation provides a powerful example of Responsible Research and Innovation (RRI). This is an approach the European Union endorses and aims to implement widely. The objective is to anticipate and assess potential implications and societal expectations with regard to research and innovation, and foster the design of inclusive and sustainable research and innovation (European Commission, Horizon 2020 Programme, 2018). We conclude this chapter with a discussion of how Patient Innovation can contribute to RRI in the healthcare sector.

4.2 PROBLEM

Health and healthcare innovation have traditionally been under the responsibility of scientists, academic institutions and pharmaceutical industry (Morrisey, 2008). However, chronic and/or rare diseases' problems and disabilities are far from being solved by the healthcare system. The needs of social providers, patients and caregivers dealing with chronic and/or rare diseases are not being met (Habicht, Oliveira and Shcherbatiuk, 2012). Traditional medicine invests in fighting disease etiology and interfering with physiopathology. However, structural damage, cognitive and physical disability, low quality of life, impairment in performing daily activities, communication difficulties and many other issues and problems afflict patients and informal caregivers in a way that motivates them to innovate and create new solutions for daily life problems. This gives room to emerging RRI in this field.

In line with that, the paradigm is changing and there is growing evidence that many patients and non-professional caregivers develop solutions to cope with their health disorders (Habicht, Oliveira and Shcherbatiuk, 2012; Kuusisto et al., 2013). Evidence also shows that these innovations have low diffusion (Oliveira et al., 2015). Too often, the innovation is not diffused beyond its developer to benefit others. The typical problems identified in the User Innovation literature (Oliveira et al., 2015) also apply to healthcare: (1) there are significant fall-offs in innovation activity as an idea progresses towards a solution in use; and (2) high-potential solutions do not reach the market and are not diffused in other ways.

4.3 CURRENT UNDERSTANDING

The healthcare sector is facing unprecedented challenges, which are magnified by increased costs, budgetary constraints, an aging population and the fairness of providing care to all (OECD, 2010). On the other hand, patients themselves are changing. They are savvier about their diseases, they expect their relation with the healthcare professionals to be open and interactive but above all they want to be part of the decision process. All of this is the reflection of what is already happening in other industries where customers have access to large amount of information (von Hippel, 1994) and became educated buyers (von Hippel, 2005; Ferguson, 2008; Anker, Reinhart and Feeley, 2011).

The rapid global diffusion of technologies has greatly improved access to knowledge. Communication is cheap, information is a commodity, and global trade increases technological diffusion. As a result, firms and users, including those outside of industrialized nations, get early exposure to the latest technologies and information. General-purpose technologies such as mobile phones and 3D printers enable individuals to solve local needs and customize products. The combined effect of these changes is having an impact on the innovation landscape.

Patients and their caregivers are able to create many different types of innovations (Zejnilović, Oliveira and Canhão, 2016). Some improve the patient's health or alleviate symptoms, others make daily life easier. Increased quality of life is sometimes the best delivery to some patients. Devices and aids developed by individuals for their own use often fall in this category.

Patient innovation and diffusion of patient's innovations are new and promising fields of study. Oliveira et al. (2015) found that only 5 per cent of patient innovators reported their innovation to medical professionals and less than 20 per cent shared it online. Continual improvements in information and communication technologies will help drive up the rates of online sharing and reporting of patient innovations.

4.4 UNMET NEEDS AND RESEARCH QUESTIONS

This is a nascent field of research and presently questions outnumber answers. Clearly more investment and research efforts are needed to develop this area of knowledge. In this chapter we will focus on three research questions: Do users (patients, caregivers and collaborators) play a role in health innovation? What are their motivations to innovate and share? Could an online open platform and social network promote the diffusion of health solutions?

In the end, using all the information, we discuss the relationship and contribution of patient innovation phenomena on RRI.

4.4.1 Objectives, Research Design and Methods

We adopted a two-stage approach to answer the above-mentioned research questions. First, we gathered evidence of Patient Innovation (PI) phenomena through online search and interviews. In the results section we describe 11 cases of innovative solutions suggested by patients themselves, their caregivers or collaborators to illustrate motivations for such innovations. Second, we describe the creation of a patient innovation web-platform and discuss its role in diffusion of patient-initiated innovation.

Step 1 – Evidence of patient innovation phenomenon

First, we searched for user-innovators and innovations in the healthcare sector. To collect as many solutions as possible we conducted an online search in the four languages dominated by the authors (Portuguese, English, Spanish and French). Several Internet engines were searched using a combination of keywords such as 'patient', 'father/mother/son/daughter', 'creates/created', 'develops/developed', 'device/therapy', 'helps/copes', for web pages that might provide references to the topic. After a first collection, we explored the potential relevance using an online in-depth assessment of worldwide newspapers, blogs, articles and social networks with an expanded set of key words. A team of three experts in the user innovation field and four medical doctors followed the definitions provided by von Hippel (2005, 2017) and Habicht, Oliveira and Shcherbatiuk (2012) to classify each solution by the type of need, relation with need, and the degree of novelty. Amongst all the discovered solutions, only the ones created to solve a personal need, a need of a relative or close friend, or by persons facing a general need in their community, were considered. We performed individual interviews worldwide with patients and caregivers. This confirmed that patient-developed innovation was indeed a reality. Data was analysed in accordance with the research design. Interviews were performed with the participants, face-to-face or by Skype, following a semi-structured questionnaire. More detailed stories were published in our platform using a template that tells who the innovator is, the background and motivation/context to develop the solution, and also describes the solution, how it was developed, how it works and its usefulness. Photos and videos accompanied the publication. The 11 cases from patients, caregivers and collaborators, with different levels of complexity show that innovation by users in healthcare is an emerging field.

Step 2 – Patient Innovation platform

With the objective of increasing the diffusion and adoption of solutions developed by patients, we established an online platform (https://patient-innovation .com). Patient Innovation is an open, non-profit, free access, multilingual online platform, founded in February 2014 in Lisbon, as a result of a research collaboration between Católica-Lisbon School of Business and Economics, MIT Sloan Business School and NOVA Medical School. This platform was designed to collect structured information about the innovators and innovations being a fruitful way to obtain data and research answers.

We gathered enough evidence of how patients, caregivers, and collaborators are important sources of healthcare-related innovations and became aware that these solutions often end up 'lost' and do not benefit other patients. As a response to this issue we created an environment to facilitate collaboration and diffusion of those innovations. We built an online platform to develop efficient, sustainable online resources for patients to research their medical questions, communicate with one another, and support each other. The Patient Innovation platform is becoming a unique repository of open-knowledge about potential healthcare solutions created and co-developed in a patient-to-patient trustful environment.

We try to optimize two key issues: (1) network effect: as more patients, caregivers, and/or collaborators share their solutions, more information will be available to those who are looking for answers to their problems and the higher the potential value is of each proposed solution, and (2) the safety character of the content shared. All the innovations submitted online are first analyzed by Patient Innovation's medical team, and only the ones that comply with the terms and services and cleared by the medical screening, are published online.

Screening of Patient Innovation's content
The screening process aims to identify and remove posts that are considered offensive or inappropriate, for commercial trade, that do not qualify as a solution proposal, that involve drugs, chemicals or biologics, that consist of non-approved by health regulatory agencies invasive devices, or that are visibly and intrinsically dangerous. However, the platform and its promoters do not scientifically or in any other way test and validate the proposed solutions.

Users of the Patient Innovation platform
We classify Patient Innovation platform's users into three categories: (1) Innovator users – those who have created a solution to cope with a health condition or disorder. Innovator users can be either a Patient; a Caregiver; or a Collaborator; (2) Active users – those who are possible adopters of the shared solutions, but who did not create any solution to cope with a health condition. They interact with the Innovator user giving feedback on the solution's performance, and (3) Visiting users – those who interact with the platform by searching and collecting information, but who do not publish directly.

The user experience and the platform design are centred on the solution's description and the reasons for the creators to innovate. Each solution, after approval by the medical team through the screening process, has a dedicated space on the platform. The aim of this strategy is to establish connections between the different actors in the community regardless of pathology. When the focus is on the solution rather than the pathology, users are exposed to the world of problem solving – small patient communities gather skills and knowledge becoming a larger patient community: the Patient Innovation community.

Community interaction

Our aim is to enable patients, caregivers, and collaborators to assume more responsibility for the collaborative development and diffusion of innovation, and to contribute more knowledgeably to their own care. Therefore, the users can interact through: (1) Solution sharing – an innovator shares his/her innovation and in return receives feedback from the community through public comments and likes; (2) Forum – this is a space dedicated to brainstorming and co-creation where a user can share a health-related question, an idea, a motivation, or a need with the community; and (3) Comments and 'Like' button – a user does not need to be an innovator to help improve lives. Users can interact with innovators by giving them a constructive opinion regarding the performance of the innovation. The 'Like' button can also be used by the user to express that they like, enjoy, or support the innovation.

4.5   RESULTS AND FINDINGS

4.5.1   Step 1 – Evidence of the Patient Innovation Phenomenon through 11 Case Studies

Our systematic online search between February 2014 and December 2017, identified 734 solutions developed by patients, informal caregivers, and collaborators. These solutions were approved by Patient Innovation medical team and published on the respective online platform (referred to December 2017). We found that patients and caregivers innovate, sometimes diffuse their innovations and occasionally become entrepreneurs. In order to give a holistic view of the spectrum of innovations found, we chose 11 cases that provide detailed insights into how patients, informal caregivers and collaborators innovate to cope with a need imposed by a health condition or disorder. Innovators from each category were chosen: patients, caregivers and collaborators. The need for a solution, the motivation to innovate, the journey pursued from the idea stage to the diffusion stage are narrated. A projected potential impact of the solution is provided in each innovator/innovation's short story.

Showering after breast surgery (patient innovator)

LC, an American broadcast journalist, was 42 years old when she was diagnosed with breast cancer. Following a mastectomy, LC was advised to avoid showering in order to prevent infection through the drain sites, since bacteria can be found in tap water. She started searching for solutions that would protect her drains. Her frustration increased when she faced the fact that there was no solution to her need. After talking to patients in a similar situation and to her doctors she realized that a significant number of individuals experienced the same problem. She decided to create her own solution to help improve the quality of life for cancer patients. What started as a plastic trash bag-based solution ended up as the first ever water-resistant garment to enable mastectomy patients to shower normally. After five different prototypes, The Shower Shirt® is patented, approved by the FDA as a Class 1 medical device, and commercially available in 36 countries. The Shower Shirt can be used by patients suffering from several conditions: chest surgery patients, including mastectomy, hemodialysis, cardiac, lung, hernia, rotator cuff, neuro-stimulation, and protect external defibrillator patients from water while showering. In its first year, around 2,200 units were sold. In the USA alone it is estimated that more than 3.1 million women live with a history of invasive breast cancer (DeSantis et al., 2014). According to this same study 36 per cent of the women diagnosed with early-stage (stage I or II) and 59 per cent with late-stage (stage III or IV) breast cancer undergo mastectomy.

Monitoring ostomy bags (patient innovator)
MS from the UK was diagnosed with Crohn's disease at the age of 12, and after several surgeries he underwent a small bowel transplant followed by an ileostomy. After this type of surgery, digestive waste passes out of the patient's body and into an external pouch and stoma bag. MS had to live with the discomfort of not being able to control the volume of the pouch which led him to isolate himself. He had to learn to monitor its amount and consistency. He could not understand why this bag has not improved to cope with this need experienced by thousands of patients. He decided to create a volume sensor using a sensor (from Wii glove) and a battery (from Blackberry). He shared his device with patients in a similar condition and with doctors to understand how he could improve the sensor. Today this device is called ostom-i™ Alert Sensor and can be attached to any ostomy bag. It sends messages via Bluetooth® to a mobile app to warn the patient when the bag is close to full. He ran a number of clinical trials across the UK, US and India. The device has been approved by the FDA as a Class 1 medical device and is commercially available through 11 Health, a firm founded by MS. The current number of ostomy patients worldwide is difficult to estimate. Around 100,000 people just in the United States undergo surgeries that result in a colostomy or ileostomy each year (Goldberg et al., 2010), while in the UK the number of people with a stoma bag is around 102,000 (Basil, 2013).

Exovasc, aortic support (patient innovator)
Being an UK engineer diagnosed with Marfan's Syndrome and a large aortic aneurism, TG realized he just had a 'plumbing problem'. In 2002 there was no solution for TG other than undergoing open heart surgery, valve replacement and aorta stent resulting in life-long anticoagulant dependency. This was his motivation to design Exovasc, an external support for the aorta root. Following TG's surgery in 2004 and being the first person to have the Exovasc inserted, another 128 patients received the same device. The surgical process is simple compared to the traditional methods of open heart surgery.

Multidisciplinary miniature 3D camera system (patient innovator)
AY from Israel, was diagnosed with an inoperable brain tumour. He built a multidisciplinary miniature 3D camera system to help in the surgery. The doctor told him no technology existed to remove such a deep-seated tumour. He was told to wait five years, and if he was lucky, someone would invent the technology. A small camera or scope that would allow the surgeon to see inside the brain in three dimensions was needed in order to successfully remove the tumour. AY took action and developed Visionsense, a miniature silicon chip that provided the necessary image-processing algorithms. The solution came too early, according to AY, who said the world wasn't ready for 3D in 2008. Ten years later the technology works and saves thousands of lives globally. Visionsense has FDA 510(k) clearance for neurosurgery.

BeMyEyes (patient innovator)
HJW suffered from tunnel vision and he knew that eventually he would become blind. The Dane is the co-founder of BeMyEyes, a company that produces a free mobile app designed to help those who are blind or visually impaired by connecting them with sighted helpers. To do this the user makes a video-call to a volunteer who can assist the visually impaired person with their task. Examples of use are navigating a new area or simply just reading a label at a supermarket. As of May 2018 BeMyEyes has registered 942,432 sighted volunteers who help 66,386 blind and visually impaired patients. It is used in more than 150 countries.

ReWalk (patient innovator)
AG is a tetraplegic engineer from Israel, who developed ReWalk robotic exo-skeleton that helps patients with lower limb disabilities to walk. ReWalk was the first commercially viable upright device that enables wheelchair users to walk and climb stairs. It is approved by FDA and is used by more than 1,000 people worldwide.

Plain white plates (caregiver innovator)
Alzheimer's disease is a degenerative brain disease and symptoms include memory loss, language impairment, and limitations in cognitive skills that affect a person's ability to perform daily activities. A man with Alzheimer's disease was having trouble feeding himself because he would try to pick up food from the rim of his dish, rather than in the centre where the food actually was. His daughter realized that the floral pattern on the rim was confusing him and switched to plain white plates. Without any patterns to distract him, her father regained autonomy and was able to feed himself just fine. This simple innovation was shared freely online. According to recent estimations, 5.7 million Americans of all ages are living with Alzheimer's dementia in 2018 (Alzheimer's Association, 2018). Simple solutions such as this one, cannot cure Alzheimer's disease but can improve patients' and their caregivers' quality of life, helping patients to regain autonomy.

Sensors for wandering (caregiver innovator)
KS's grandfather from USA suffers from Alzheimer's disease and wanders alone at night. KS, a technology-minded teenager, created SafeWander™, a thin, flexible pressure sensor that attaches to any piece of clothing and alerts the caregiver, via Bluetooth® and a smartphone app, when the patient gets out of bed. KS has patented his invention, which has won multiple awards, and has created a company to market it. According to the Alzheimer's Association, non-professional unpaid caregivers count for around 50 per cent of the help provided to Alzheimer's patients. Around 34 per cent is over 65 years old. When compared with caregivers of people without dementia, twice as many caregivers of those with dementia indicate substantial emotional, financial and physical difficulties.

FasoSoap (caregiver innovator)
GN saw three of his siblings die of malaria, and GN himself suffered from near-fatal malaria. GN lives in a poverty stricken area of Burkina Faso where incidence of malaria fatalities is high. This spurred GN to develop a low-cost method to fight malaria. He created a soap made of ingredients extracted from local plants. All the ingredients are natural and locally available in Burkina Faso. This solution, added to locally manufactured soap, provides a very accessible, low-cost anti-malarial protection.

The soap makes the mosquitos stay away for six hours and is an affordable solution for all families in developing countries. In an interview, GN stated that the company's goal is to save 100,000 lives in the Democratic Republic of the Congo, Tanzania, Ghana, Nigeria and Uganda through malaria prevention by 2020.

3D-printed arm and hand for children (collaborator innovator)

Ivan is an artist who posted a video on YouTube about a mechanical hand he had created. A carpenter who had lost some fingers saw the video and contacted Ivan asking him to design a prosthetic hand. Ivan accepted the challenge and he now makes low-cost 3D printed hands.

The design of these gadgets is open source. As a result, e-NABLE, a global network of volunteers who use 3D printing to give the world a helping hand, was established. The e-NABLE Community started with around 100 people that offer 3D printing devices for free. Within its first year the e-NABLE community grew from 100 members to 3,000+ and its members made around 750 hands. The following year they nearly doubled membership to 7,000 and approximately 2,000 devices were created and gifted to individuals in over 45 countries. e-NABLE membership is currently growing by a few per cent each week.

Fold-up wheels (collaborator innovator)

When he was a graduate student, DF from UK invented folding wheels for bicycles. He was exhibiting them at a bike show when a wheelchair user approached and asked if he could adapt the technology for wheelchairs, to make them easier to take on a plane. The resulting product, Morph Wheels, transforms a wheelchair into a small, neat package that can go into an overhead bin or the trunk of a car. Morph Wheels is patented, and distributed by 7th Design & Invention, a product design studio that DF co-founded.

4.5.2   Step 2 – Patient Innovation Platform

Patient innovation platform content assessment

From February 2014 until December 2017, the platform achieved a considerable growth rate of around 10 per cent per month on average, resulting in 734 solutions published online on www.patient-innovation.com (considering only those who successfully passed the medical screening) of 55 countries. The top five contributing countries are: United States of America (49 per cent), United Kingdom (11.10 per cent), Belgium (5 per cent), Canada (3.75 per cent) and Australia (3.5 per cent). Regarding the type of innovator, 43 per cent of these solutions were developed by patients, 40 per cent by caregivers, and 17 per cent by collaborators. Two analysts classified 95.2 per cent of these solutions as products, 4.1 per cent as strategies and/or tips, and 0.7 per cent as services. Additionally, the type of product was explored, and 82.4 per cent of the products were classified as devices, 13.2 per cent as software (mobile apps included), and 4.4 per cent as hybrid (a hybrid solution is a combination of a software and a physical device).

Around 57 per cent of the innovators created a solution because the products in the market did not satisfy the need, while 33 per cent of innovators created solutions because there was no alternative in the market. Approximately, 10 per cent of the innovators developed a low-cost solution to overcome the cost of the alternatives in the market.

Regarding the stage of development, 1.5 per cent are in idea stage, 37.2 per cent are functional prototypes and 61.3 per cent are final products.

In order to understand to what extent solutions developed by patients and informal caregivers would require clearance or approval from health regulatory agencies we attempted to classify all solutions (even the non-commercial ones), according to the Code of Federal Regulation (CFD), into the three categories established by the FDA (Foods & Drugs Administration) based on the level of risk posed to a patient.

In the case of the non-commercial solutions we followed the FDA guidelines for medical device classification. There are two strategies to determine the classification: first, we searched the FDA databases for a substantial equivalent device and when found, we extracted the regulation code and classification name. Second, we identified the device panel (medical specialty) to which the device belongs, identified the general family of the device and the corresponding regulation classification. Half of the solutions are or would be considered medical devices and therefore, would require health regulation screening. However, only 13 per cent of the solutions are actually cleared or approved by FDA, and 2 per cent are waiting for approval. According to FDA, the classification of a device is based on the level of risk imposed to the patient, its intended use and technological characteristics. From our classification, we found out that 64.1 per cent of the solutions were classified as device class I (low risk, which requires least amount of regulatory control), 29.5 per cent device were class II (moderate risk, it requires special controls to assure safety and effectiveness), 1.1 per cent class III (high risk, it requires premarket approval since they are intended to support or sustain human life), and 0.8 per cent of the solutions would be or were classified as De Novo devices, which means that this type of device was not previously classified. Ultimately, 4.5 per cent are classified as software as medical device.

Concerning the type of diffusion channel chosen by the innovators, we found that 50.1 per cent were diffused via commercialization, of which 55.2 per cent requires FDA approval, and 49.9 per cent were freely shared, of which 45.7 per cent requires FDA approval.

Patient innovation platform diffusion
In terms of acquisitions, approximately 63 345 users have accessed the Patient Innovation platform (on average 1319 users per month). Approximately, 58 per cent of the users are male and 42 per cent are female. The 25–34 age group

represents 34 per cent of user traffic, users in the 45+ age group represent 28 per cent, followed by the 35–44 age group (25 per cent), and the 18–24 age group (13 per cent).

Regarding the origin of Patient Innovation's users, the top five countries are: Portugal (33.5 per cent), Brazil (15.7 per cent), US (14.6 per cent), Germany (5.3 per cent), and UK (4.8 per cent). The majority of the cases (60 per cent) users access Patient Innovation platform through organic search (natural search), which is a search that generates results that were not paid advertisements. This highlights that there exists a deep need among patient innovators for sharing and diffusing their ideas, and searching for solutions developed by others with similar needs.

4.6 CONTRIBUTION

From hacking medical devices, designing technical aids from scratch, and finding new therapies, developing technological breakthroughs to rebuild body parts, patients and caregivers are shaping the healthcare of the future. Patient innovation can potentially transform healthcare the way open-source software has transformed technology. We are just at the beginning of understanding how and why patients innovate, and how to integrate those innovations to benefit the entire system. The concept of responsible research and innovation by the European Commission "implies that societal actors (researchers, citizens, policy makers, business, third sector organizations, etc.) work together during the whole research and innovation process in order to better align both the process and its outcomes with the values, needs and expectations of society" (European Commission, Horizon 2020 Programme, 2018). In fact, all stakeholders should be engaged in research and innovation, enabling easier access to scientific results and actively participating, regardless of having formal or informal science education. Health is a matter of interest for all citizens, not just healthcare professionals or policy makers; it is of interest to all patients and informal caregivers, who can be anyone. Thus, RRI in healthcare is mandatory, to achieve better outcomes and sustainability.

But, numerous challenges remain, and it is vital that all stakeholders learn to recognize and nurture innovation whenever and wherever they find it. In this work, using a double stepping strategy, we first identified individual cases of patients and caregivers innovating to solve their own health problems and disabilities. We then presented our effort in establishing an open and free online platform where patients, caregivers and collaborators interact, helping each other, sharing solutions, bringing new solutions to the market and to society, contributing to increase awareness and ultimately changing healthcare and social systems.

Future research can be built upon this platform to test the efficacy, safety, cost-effectiveness and impact of a variety of interventions aimed at facilitating collaboration and diffusion of Patient Innovation solutions. Understanding the features that promote community interaction in an innovation environment can be used to identify the drivers for diffusion and adoption of user innovations.

REFERENCES

Alzheimer's Association (2018). What is Alzheimers. Retrieved 1 March 2019 from https://www.alz.org/alzheimers-dementia/what-is-alzheimers.

Anker, A. E., Reinhart, A. M. & Feeley, T. H. (2011). Health information seeking: a review of measures and methods. *Patient Education and Counseling*, 82(3), 346–354. https://doi.org/10.1016/j.pec.2010.12.008.

Basil, N. (2013). Stoma care: the market in products lets patients down. *British Medical Journal*, 347, 6129. https://doi.org/10.1136/bmj.f6129.

DeSantis, C. E., Lin, C. C., Mariotto, A. B., Siegel, R. L., Stein, K. D., Kramer, J. L. & Jemal, A. (2014). Cancer treatment and survivorship statistics. *CA: A Cancer Journal for Clinicians*, 64(4), 252–271.

European Commission, Horizon 2020 Programme (2018). Responsible research and innovation – Horizon 2020 – European Commission. 11 January. Retrieved 1 November 2018 from https://ec.europa.eu/programmes/horizon2020/en/h2020-section/responsible-research-innovation.

Ferguson, T. (2008). E-patient: how they can help us heal healthcare. In J. A. L. Earp, E. A. French and M. B. Gilkey (eds), *Patient Advocacy for Health Care Quality: Strategies for Achieving Patient-Centered Care* (pp. 93–120). Sudbury, MA: Jones & Bartlett Learning.

Goldberg, M., Aukett, L. K., Carmel, J., Fellows, J. & Pittman, J. (2010). Management of the patient with a fecal ostomy: best practice guideline for clinicians.. *Journal of Wound Ostomy & Continence Nursing*, 37(6), 596–598.

Habicht, H., Oliveira, P. & Shcherbatiuk, V. (2012). User innovators: when patients set out to help themselves and end up helping many. *Die Unternehmung*, 66(3), 277–294.

Kuusisto, J., De Jong, J. P., Gault, F., Raasch, C. & von Hippel, E. (2013). *Consumer innovation in Finland. Incidence, Diffusion and Policy Implications*. (Proceedings of the University of Vaasa, 189). Retrieved 1 March 2019 from https://www.researchgate.net/publication/262115942_Consumer_Innovation_in_Finland_Incidence_Diffusion_and_Policy_Implications.

Morrisey, M. (2008). Health care. *The Concise Encyclopedia of Economics*. Retreived 15 July 2014 from http://econlib.org/library/Enc/HealthCare.html.

OECD. (2010). Innovation to strengthen growth and address global and social challenges. Retrieved 28 November 2013 from http://www.oecd.org/sti/45326349.pdf+.

Oliveira, P., Zejnilovic, L., Canhão, H. & von Hippel, E. (2015). Patient innovation under rare diseases and chronic condition. *Orphanet Journal of Rare Diseases*, 10(1), 41. doi: 10.1186/s13023-015-0257-2.

von Hippel, E. (1994). 'Sticky information' and the locus of problem solving: implications for innovation. *Management Science*, 40(4), 429–439. https://doi.org/10.1287/mnsc.40.4.429.

von Hippel, E. (2005). *Democratizing Innovation*. Cambridge, MA: MIT Press.

von Hippel, Eric (2017). *Free Innovation*. Cambridge, MA: MIT Press.

Zejnilović, L., Oliveira, P. & Canhão, H. (2016). Innovations by and for patients, and their place in the future health care system. In H. Albach, H. Meffert, A. Pinkwart, R. Reichwald and W. von Eiff (eds), *Boundaryless Hospital* (pp. 341–357). Berlin, Heidelberg: Springer.

5. Patient-initiated innovation – evidence and research agenda

Thomas Laudal and Tatiana Iakovleva

5.1 INTRODUCTION

Historically, the doctor–patient relationship has been based on a one-way trust. Hippocrates advised doctors to conceal most things from patients in order to keep them calm and to ensure medical authority.[1]The doctor–patient relationship is still basically a one-way communication relationship (Epstein et al., 1993), sometimes referred to as a 'paternalistic relationship' (Charles, Gafni and Whelan, 1999). However, since the establishment of professionalized healthcare, there have been calls for improving the dialogue between the healthcare provider and the patient. Katz (1984) states there is a need for an informed dialogue between patients and doctors in order to respect the rights and needs of both sides. Eric Topol refers to the continuing advances in ICT when he states that:

> The percentage of people who have and understand their own medical data is in the single digits, but there's the new potential for information parity and equality and, as a result, emerging authority for *you* to be the decision maker. (Topol, 2015, p. 14)

Eric Topol argues that the potential for information parity and equality between patients and healthcare providers is linked to advances in the general education level, improved information access, and to the increasing processing power of our computers. Information parity and equality indicates that there have been two changes in the status quo: That patients' knowledge of their health condition has improved, and the contact between patients and healthcare providers is more balanced – leaving more room for patients to participate in their treatment. According to Teodoro (2016, p. 3) the need to rethink the traditional conception of patients as consumers of health communication appears to be justified by a number of research insights linked to social psychology (increasingly rich stimuli-response environment), differences in patients' ability to take advantage of health information, and different contexts influencing patients' ability to make correct decisions.

This is central in what is often referred to as a 'patient centred care'. Patient centred care has been common in most western countries for some time (Stewart, 2001; Hernandez et al., 2013; Cleary, 2016). The literature covering this trend emphasizes the importance of adapting and listening to the patient in different stages of the treatment. However, it does not refer to the feedback patients give as a potential for institutional innovation. The concept 'Institutional innovation' is discussed by many scholars. Hargrave and Van de Ven (2006, p. 866) defines 'institutional *change*' as a difference in form, quality, or state over time, in an institution. Changes can be determined by studying the frames, norms and rules in institutions. And if the change is a novel, or unprecedented departure from the past, then it represents an 'institutional *innovation*' (Hargrave and Van de Ven, 2006). Hagel and Brown (2013, p. 4) define 'institutional innovation' as a change that "redefines the rationale for the institution and developing new relationship architectures within and across institutions". In this chapter, institutional innovation refers to the less demanding definition of Hargrave and Van de Ven (2006). But the improved access to health information and the facilitation of the dialogue between patients and healthcare providers through new ICT systems, suggests there is a potential for patient-initiated innovations affecting hospital performance.

Patients[2] are better informed than ever about their medical and procedural interests due to the improving access and proliferation of digitalized information. It is reasonable to expect that patients will continue to utilize this information in their communications with healthcare providers. A well-functioning communication system, allowing clinicians to consider follow-up actions based on patient feedback, may improve the performance of the hospital. Patients may highlight issues that medical staff are not aware of, or they may highlight how a healthcare procedure may be improved. Thus, patients may initiate institutional innovations. This is the assumption that motivates the proposal for a research agenda in this text.

The aim of this chapter is threefold: to present a preliminary review of the literature on institutional innovations linked to hospital patients, to delineate the main concepts we need in this research area, and to present a research agenda to better understand how patient feedback may enhance hospital performance.

First, we will describe technology trends suggesting that there may be a potential for more patient-initiated innovations in the healthcare sector. After this we present how electronic health records (EHRs) has been disseminated to patients in the last decade. This is a significant development because EHRs are an important source for patients seeking to understand and participate in treatments. We then present our two expectations based on these trends: an increase in the number of feedbacks from patients, and an increase in institutional innovations in hospitals. The first expectation is related to the release of EHRs and increased access to relevant health information in general. The second is based on the assumption that a proportion of these feedbacks will include proposals for change. We then consider the literature on how patients may contribute to institutional innovations. To illustrate these issues, we present examples of patient feedback based on interviews in a large regional hospital in Norway. We then consider how patient feedback may lead to innovations that enhance hospital performance. Characteristics of responsible innovation are then presented as they may be helpful in validating patient-initiated innovations. Finally, we present some of the problems and questions we believe are important in future empirical studies in this field. The contents of this chapter are summarized in Figure 5.1.

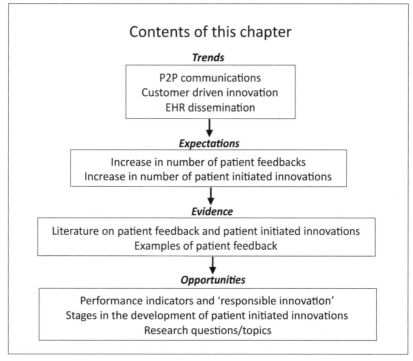

Note: Arrows are not to be interpreted as causal relations, only as the order of topics in this text.

Figure 5.1   Contents of this chapter illustrated

5.2   PEER-TO-PEER TECHNOLOGIES AND CUSTOMER-DRIVEN INNOVATIONS

Scholars refer to two trends that link new communication structures with new innovation patterns; peer-to-peer technologies, and customer-driven innovations. Peer-to-peer communication refer to networks where the nodes (individual actors) communicate without third parties as brokers or mediators. A growing fraction of our world's economy consists of digitally enabled peer-to-peer exchange (Oram, 2001; Cohen and Sundararajan, 2015). The next global communication standard developed by ETSI (GSM), broadly referred to as '5G', is to a large extent based on peer-to-peer communications (Bangerter et al., 2014) and will substitute significant parts of fixed line broadband communications (Blackman and Forge, 2016). Finally, new network technology is allowing document validation referred to as 'block chain' or 'distributed ledger technologies' (Xu et al., 2017) and 'tangle' (Popov, 2016) to enable peer-to-peer communication and machine-to-machine communications.

The second trend is 'customer-driven innovation' (Thomke and von Hippel, 2002; Desouza et al., 2008). This refers to the tendency to invite corporate suppliers/customers in as developers and allowing them to determine adaptations to prototypes. This is often associated with customization and the ability to adjust business models to fit low-volume customers. When customer-driven innovation concerns the implementation of ICT applications in large organizations it is often referred to as 'agile project management'. This is characterized by mutual interactions among a project's various parts in the direction of continuous learning and adaptation (Augustine et al., 2005, p. 87). There are several examples of agile project management when ICT systems are introduced in hospitals (Laudal, Bjaalid and Mikkelsen, 2016).

The trends towards peer-to-peer innovations and customer-driven innovations suggest there is a potential for more patient-initiated innovations when patients get access to their electronic health records (EHR) as a part of an ever increasing array of openly available medical information.

5.3   DISSEMINATING ELECTRONIC HEALTH RECORDS TO PATIENTS

According to Bos et al. (2008) several surveys show that a majority of patients searched for health information on the Internet and discussed this with their doctor. To achieve true patient empowerment a bi-directional contact between patients and clinicians is necessary (Bos et al. 2008).

Healthcare institutions across Europe and the US are not only disseminating EHRs[3] among healthcare professions, and developing standards to allow the exchange of EHRs among healthcare institutions. Patients are being included in the loop. Most EHR applications offer a patient-facing interface, commonly referred to as a Patient Portal (Garrido, Raymond and Wheatley, 2016). These applications generally allow patients with Internet connectivity and computer access to view portions of their medical records. Not all health institutions have adopted EHRs, but most save their patient information in a digital format. Allowing patients to access their EHRs is not sufficient. Patients must be able to interpret their EHR and apply this in their own decision-making processes, according to Howrey et al. (2015).

The US Federal Government (Office of the National Coordinator for Health Information Technology) reported that 96 per cent of non-Federal acute care hospitals had adopted a Certified EHR by 2015 (ONC, 2015).

In the US 83 per cent of US hospitals had implemented at least a basic EHR in 2015 (ONC, 2015). A survey conducted by HealthMine (2016) showed that 60 per cent of the population using Internet-connected applications in 2016, had access to an EHR. However, the vast majority of states in the US have only instituted the most rudimentary of exchanges. Interoperability of EHRs and cost are frequently-cited barriers to broader implementation.

In the UK public access to a 'summary care record' was in place in 2016 (NHS, 2016). The national goal is that the EHRs in the NHS, including pharmacies, should be fully digitalized and standardized in 2020 (House of Commons, 2016). In the Conservative Party Manifesto for the 2017 elections (2017, pp. 68–69), patients are promised "… the ability to … access and update aspects of their care records, as well as control how their personal data is used". Access to EHRs across secondary care units in England were developed and implemented far more slowly than was originally envisioned (Robertson et al., 2010). This was due to political and financial factors. Empirical studies record strong support for EHR implementation at the hospital level.

EHRs has been part of the standard documentation and communication tools in public hospitals all over Scandinavia since 2005–2010. In Denmark, online dissemination of EHRs to patients started in 2009 (Høj, 2009) and in Sweden in 2012 (Huvila, Myreteg and Cajander, 2013). In Norway this process started in 2015 as part of a national programme managed by the Norwegian Directorate of eHealth (2016).

For EHRs to be useful for patients, and their close relatives and friends, one needs to have access to the Internet. According to Eurostat (2018) the proportion of regular Internet users (having used the Internet in the last three months) in the age group 65–74 years in the EU has risen from 19 per cent in 2008 to 52 per cent in 2017. We find the highest proportion of regular Internet users in all age groups in Norway and Iceland (98 per cent). Among those in the age group 65–74, we find the highest proportion of regular Internet users in Northern Europe. Thus, the proportion of regular Internet users among the older cohorts is growing fast, and it is particularly high in Northern Europe. This indicates that 'access to Internet' is not a barrier for patients wanting to access their EHR in this region.

5.4 EXPECTATIONS RELATED TO PATIENT FEEDBACK AND INNOVATIONS

Based on the increasing use of the Internet among all age groups, the dissemination of EHRs to patients, and the trends towards peer-to-peer technology and customer-driven innovations, two expectations seem reasonable:

- *Hospitals will receive more feedback from patients.* We would expect a growing number of feedbacks from patients, and close relatives and friends, to hospitals. These feedbacks will be based on patients' experience of symptoms, their interpretation of health information, or their interpretation of agreed schedules and treatment procedures.
- *Patients will start contributing to institutional innovations.* Given that a proportion of all patient feedback will be in the form of an opinion proposing some kind of change, we expect that a growing number of feedback from patients will result in a growing number of proposals for change. And given that a proportion of patients' proposals for change are well-founded and will be welcomed by the hospital staff, we also expect that more patient feedback will contribute to patient-initiated institutional innovations.

There is an important distinction between patients contributing to hospital innovations and patients challenging the authority of medical professions. The promise of empowerment in patient-related ICT projects is normally about giving patients opportunities to participate and learn, not about challenging the authority of medical professions. Lupton and Jutel (2015) finds that many apps offering self-diagnosis blur this distinction. ICT firms may have commercial interests in exaggerating the role of these apps. In this text we focus on interactions between patients and healthcare institutions and do not consider the role of commercial apps.

If we will see a growing number of institutional innovations being triggered by feedback from patients, how are hospitals to ensure that these are implemented in a manner that maximizes the benefits for patients? This is where the literature on 'responsible innovations' may be helpful (Owen et al., 2013). First we will take a closer look at how patient feedback may lead to institutional innovations, according to the literature. Then we may consider how the characteristics of a responsible innovation can help us to judge whether a patient-initiated innovation enhances the performance of hospitals.

5.5   THE LITERATURE ON PATIENT FEEDBACK AND HOSPITAL INNOVATIONS

A literature review based on a group of selected journals was conducted in Google Scholar (https://scholar.google.com) for contributions published in 2015 or later.[4] We selected four relevant subcategories in the category 'Social Sciences', and two relevant subcategories in the category 'Business, economics and Management' in Google Scholar. Within each of the six subcategories the 20 most cited journals were considered. The 14 most relevant journals based on their titles were then selected. In addition three journals were added based on a search for journal titles that included the word 'innovation'.

A Google scholar search for four keywords (anywhere in the text) was then conducted covering the 17 journals.[5] The keywords were 'patients', 'innovation', 'hospital', and 'digital'. The search resulted in 129 articles,[6] as shown in Table 5.1.

Exclusion criteria were then applied in an individual review of each article:

- Articles concerning contexts other than hospitals, e.g., primary healthcare or national healthcare policies were excluded.
- Articles that are not including patients' contribution to healthcare performance were excluded.
- Clinical studies concerning specific medical treatments were excluded.

This screening reduced the original group of 129 articles to nine. Seven articles from seven different journals were added, based on references in the nine articles.

5.6   RESULTS OF THE GOOGLE SCHOLAR LITERATURE SEARCH

Ledford, Cafferty and Russell (2015, p. 77) refer to studies showing that:

> Patients seek information to understand a diagnosis, to decide on a particular course of treatments, and to help make prevention decisions. Patients also take an increasingly active role in medical decision-making and increasingly seek health information outside of the medical encounter.

Not surprisingly, Ledford et al. find that the most 'health literate' patients are the ones most eager to search for information about their illnesses. To fully utilize the patient's potential to help make the diagnosis and determine the best treatment, Ledford et al. recommends that healthcare providers ask patients if they have obtained relevant information on their own at an early stage.

Table 5.1   *The sample of journals included in the Google Scholar search*

Journals	h5-index*	Subcategory / Source	Articles
R&D Management	31	Entrepreneurship & Innovation	1
Organization Studies	52	Human resources & Organization	1
American Journal of Public Health	86	Public Health	5
European Journal of Public Health	48	Public Health	18
Journal of Health Communication	37	Public Health	4
Public Administration Review	51	Public Policy & Administration	3
Public Management Review	41	Public Policy & Administration	10
Social Science & Medicine	75	Social sciences (general)	14
Harvard Business Review	63	Strategic management	4
*Journal of Management***		Strategic management	66
Creativity and Innovation Management	29	*Search for 'innovation' in journal title*	1
International Journal of Innovation Management	22	*Search for 'innovation' in journal title*	1
Journal of Responsible Innovation	19	*Search for 'innovation' in journal title*	1

Notes:
Articles are search results in these journal including the four keywords (anywhere in the text): 'patients', 'innovation', 'hospital', and 'digital'.
* The h5 index shows the *maximum* number satisfying the requirement given that 'h'= 'the number articles the last completed five years' *and* 'h'='the minimum number references in any of these articles'.
** '*Journal of Management*' includes all journals including this phrase in their title; e.g., 'Journal of management information systems'.

The effects of being an engaged patient is studied by Judith H. Hibbard. She focused on the effects of patients being more or less engaged in, or knowledgeable of, the health issues and treatment they receive. In 2004 Hibbard published a validated survey instrument for measuring the degree of patient engagement; the Patient Activation Measure (PAM)[7] (Hibbard et al., 2004). In the years following, Hibbard and her colleagues showed that PAM predicted a wide range of health behaviours (Hibbard and Mahoney, 2010 and Greene and Hibbard, 2012). In recent years, studies show how interventions may stimulate the patients' engagement (increasing their PAM score) and contribute to increased innovation in medical units (Hibbard and Greene, 2013). Several indicators of PAM are related to the patients' perceived autonomy. This is measured by two survey items 'I know about the self-treatment for my condition' and 'I know how to prevent further problems with my health condition'. This perception is similar to the experience patients have when they access their medical records online, according to Hibbard et al. (2004).

Researchers in Boston, Schnipper et al. (2008), reported the deployment of a web-based patient portal linked to an Affordable Care Act ambulatory care in 2006. Results showed that usage and satisfaction data indicate that patients:

- felt that it led to their providers having more accurate information about them, and
- made them feel more prepared for their forthcoming visits.

This is in line with interviews conducted by Honeyman, Cox and Fisher (2005) where patients were more interested in the EHR than in their paper record because they trusted the EHR more. More recently Ahmed et al. (2016) reported their results of a randomized clinical trial assessing the effectiveness of a novel web-based Asthma Self-Management System. They found that for all self-reported measures, the intervention group had a significantly higher proportion of individuals, demonstrating a minimal clinically meaningful improvement compared with the usual care group.

Thus, there are several studies showing how patients benefit from being active and informed during their treatment. However, very few studies refer to how the active and communicating patients may also benefit the performance of the hospital. This may partially be explained by the predominance of the current 'patient-centred care' paradigm. According to this paradigm, patients are both the object of care, and the source of its validation. According to Stewart (2001), the patients' opinions should be emphasized because it has been shown that this has a significant impact on the health outcome of the individual patient. However, in Stewart (2001) and other literature in this field, the patient's opinion is not valued as an input to improve the group of patients with issues relevant to the one being voiced by the initial patient. Similarly, in a literature review of the effects of public reporting of health institutions' performance data, we see that the considered effects for patients were restricted to those influencing their individual health outcome (Vukovic et al., 2017, p. 976). No study considered how public reporting could affect patients' motivation and capacity to institute institutional innovations. An equally restricted perspective on patients' engagement is found in a study of personalized healthcare (Ricciardi and Boccia, 2017), on the effects of introducing EHR (Campanella et al., 2015; Cucciniello, et al., 2015), and on a redesign of hospital pharmacy services (Lindsay et al., 2018).

Among general practitioners (GPs) we find evidence that patients have been regarded as possible sources of innovation. In a qualitative study in the Netherlands of 63 GPs and medical students, it was found that healthcare providers were guided by five goal categories (Veldhuijzen et al., 2011). Two of these were labelled 'maintaining organization of care' and 'maintaining public health agenda'. These categories included the operational goals 'time management' and 'reducing healthcare costs'. This shows that the GPs were not only focusing on the health conditions of their individual patient, but also on the potential for institutional innovations linked to their patient contact. But the study did not document that the GPs consciously tried to utilize patient feedback as input in innovation processes.

This literature survey shows that most research covering the communications between patient and healthcare institution focuses on how this communication can benefit the individual patient. The topic of how patient feedback may lead to institutional innovations, is very rarely covered in the literature. One reason for this could be that hospitals do not receive much feedback from patients of the sort that inspires institutional innovation. However, this is not a likely explanation, given the many examples of feedback from hospital practices.

5.7 EXAMPLES OF PATIENT FEEDBACK FROM A LARGE REGIONAL HOSPITAL IN NORWAY

Based on ten interviews of employees at a large regional hospital in Norway,[8] we found that patient feedback was widespread throughout the hospital. This very small sample of hospital employees made reference to nine incidents where patients' feedback was considered potentially valuable. Six of the feedbacks were acted on and corrected, but none of the actions were implemented through a standard procedure or a system designed to support the follow-up actions of employees.

5.7.1 Feedback from Patients

- Occupational therapy in a rehabilitation unit:
  - Patients expressed a wish to reduce time spent on an introductory meeting. A six hour meeting was then reduced to four hours. The schedule had to be compressed.
  - Patients complained that they were cold at meetings because they were scheduled after a joint training session where many were perspiring. The department changed the order so every meeting was before training sessions.
- Psychiatric polyclinic:
  - The hospital unit's message to the patient about the next appointment, and the message-box available for the patient to submit comments to the unit only allow for one-way communication. Patients have complained

about this, but the unit does not see how this can be transformed to two-way communication systems. Two-way communications requires patients calling them on a telephone which is often busy.

- Several patients were critical of the design of the waiting room. The restroom was adjacent to the waiting room and the waiting room is an open space which does not allow for privacy in any way. The unit does not see that they have any means to change the room layout in this building and our informant did not know of any initiative to change the layout of the waiting room.
- Parents of a patient complained that a serious diagnosis they had received from the clinic was not well explained. The clinic decided to be more aware of this challenge and started to have systematic feedback meetings with patients when they were done with treatment.
- Rehabilitation function in a psychiatric unit:
  - Relatives of patients were contacted by telephone but thought that it was difficult to explain their concerns in this medium. The unit decided to offer the relatives a face-to-face meeting in all similar situations.
- Physiotherapy function in a children department:
  - Some parents of the children patients are reading their child's EHR and ask the unit to make changes in cases where they don't agree, or if they believe there is a mistake. However, the unit suspects that many parents are not aware that they have access to their children's EHR. No action to better inform dependents was reported.
- Physiotherapy function in an observation unit:
  - An adult patient born with a specific lung condition was admitted to the unit and knew very well what medicine the condition required. However, the unit could not give this medicine without the paper-based discharge summaries from another hospital. This caused much irritation. According to the unit they have little influence on the ICT systems and have therefore not taken action in this case.
- Learning and coping unit:
  - Experienced patients have insisted on introducing two new themes in one of the courses for former patients. The unit followed their advice and included these themes in later courses.

In general, many patients attending courses in the learning and coping unit discuss experiences they disapprove. Course lecturers (medical professions from other departments) report to the unit that meeting patients in a group setting is very different from the one-on-one meetings they are used to.

This list of feedback based on ten interviews at this hospital indicates that there is no lack of feedback from patients. The challenge is how to record these feedbacks, evaluate them, and (possibly) act on them. There are no systems or routines implemented at the hospital level, or at the second level of the hospital, to take on these functions.

Before we propose a research agenda in this field, we should consider how patient-initiated innovations may impact hospital performance.

5.8 THE LINK BETWEEN PATIENT-INITIATED INNOVATIONS AND HOSPITAL PERFORMANCE

When we observe an intended institutional change in hospitals this will normally be referred to as an *improvement* by those in charge of this change process. To fulfil the criteria of Joseph Schumpeter's broad innovation categories (Schumpeter, 1983),[9] we need to qualify this change by demanding that the change should be categorized as an improvement of the *hospital's performance*. The changes we are studying here are in line with the concept 'expertise-field innovation', presented by Ina Drejer (2004). Expert-field innovations is associated with "detecting new needs" and "responding to them through a procedure of accumulating knowledge and expertise ... in an interaction with a client" (Drejer, 2004, p. 559).

As we focus on innovations originating from *patients*, implying that the innovation should lead to a tangible improvement in hospital performance, we may further qualify what we mean by an innovation: The innovation should be conceived of as an improvement *from the patients' perspective*. According to Simula (2007) the 'goodness' of an innovation is a topic that is not studied thoroughly. This is due to the classical assumption that an innovation should be a lasting change in a product, or a process, that improves productivity and thereby improves the economic performance of an organization. Thus, 'goodness', in the meaning of improving the *economic* performance of business, is *implied* in the definition of an innovation (Simula, 2007). However, innovations in hospitals, understood as changes that improve hospital performance, do not necessarily imply an improved economic outcome. Qualified innovations do not even guarantee an improvement in any formal performance indicator. Innovations in hospitals refer to a wide range of performance indicators (Copnell et al., 2009; Klazinga, Fischer and Ten Asbroek, 2011), and it is an ongoing discussion whether the current indicators can capture the range and complexity of health service activity that represent an improvement from the patients' perspective (Davies and Lampel, 1998, Carinci et al., 2015, Austin et al., 2015).

Thus, to validate whether the dissemination of EHR information to patients may lead to patient-initiated innovation, an improved economic performance or a tangible impact on a formal performance indicator is not required. It would not be surprising if some of the innovations resulting from patient feedbacks are related to procedures or treatments that are *not* identified as a formal performance indicator. This could even explain why there is room for improvement in this area. Instead of relying on formal performance indicators we may validate patient-initiated innovations by treating the characteristics for 'responsible innovations' (RI) as qualifying characteristics. Innovations fulfilling these characteristics (adapted to our context), we assume are benefiting the targeted patient group the most. Thus, if a patient-initiated innovation fulfils the characteristics of a RI, we assume the innovation contributes to hospital performance in the best way. This follows from the virtues of the four characteristics of RIs: Those in charge of the innovation must ensure that both long- and short-term impacts are taken into account, they must reflect on their own role throughout the process, they should make sure that all stakeholder interests are taken into account, and they should reflect on whether the initial objectives of the innovation project still hold throughout the process.

We need to take a closer look at the characteristics of RI to see how this may help us to validate patient-initiated innovations: We refer to four characteristics of responsible innovations as described in Owen et al. (2013). According to Owen et al. (2013) responsible innovations should entail a collective and continuous commitment in four areas (possible indicators in a hospital environment are added by the authors):

1. ***Anticipatory:*** How far have we thought through the future implications of what we are trying to do? What implications will the dissemination of EHRs to patients have for different user groups?
   Possible indicator in a hospital environment: General applicability and future implications is discussed in at least one meeting among healthcare providers.

2. ***Reflective:*** How far do we challenge ourselves, discuss and debate what we are trying to do, look at it from different perspective? Do hospitals reflect on their current practices?
   Possible indicator in a hospital environment: The issue of whether follow-up actions actually will benefit a *category* of patients, is discussed in at least one meeting among healthcare providers.

3. ***Deliberative:*** How far do we try to include the views and perspectives of our planned end users and other possible stakeholders? Do professional healthcare providers (nurses, doctors, others) seek information from patients?
   Possible indicator in a hospital environment: Hospital employees receiving feedback from the patients have contacted these patients again and exchanged views on possible follow-up actions.

4. ***Responsive:*** Does our design allow us to change our response to the above questions if our initial assumptions do not hold? Is the hospital able to

adapt their strategy/actions in response to new information or new view-points expressed by stakeholders?

Possible indicator in a hospital environment: The issue of having to adapt to unforeseen implications linked to a follow-up of patient feedback has been discussed in at least one meeting among healthcare providers.

RI is here conceived of as qualifying characteristics of hospital innovations. To fulfil the criteria of innovation in order to 'contribute to something good' (Simula, 2007), we contend that the characteristics of RI are useful for deciding whether changes based on patient feedback qualify as an innovation. This is parallel to how Hernandez et al. (2013) argue that the eight principles of 'patient-centred care'[10] amount to defining characteristics of 'patient-centred innovation', and even to the characteristics of the more general concept 'organizational innovation'.

An important aim of the research agenda we propose here is to consider the actual relationship between patient-initiated innovations fulfilling the characteristics of RI, and institutional performance. In line with the expectations in Owen et al. (2013) we hypothesize that there is a positive relationship. We should determine whether the innovation linked to a particular patient feedback impacts the performance of the hospital. We propose the following two indicators of such an impact:

- Empirical examination shows that additional patients to the one initiating the innovation (giving feedback) will benefit from the change.
- The hospital has institutionalized the change. That is; hospital norms, procedures, and/or medical treatments are changed to be in line with the changes suggested in the patient feedback.

The relationship between patient feedback, responsible innovations, and institutional performance is illustrated in Figure 5.2.

5.9   STAGES IN THE DEVELOPMENT OF PATIENT-INITIATED INNOVATIONS

We envisage different stages in the development of patient feedbacks and institutional innovations related to these feedbacks. Empirical research on hospital innovations may refer to these stages. The stages may not appear sequentially, or in the order listed in Figure 5.2. The purpose here is to map the stages involved in developing patient-initiated innovations. Figure 5.2 may be used as a reference when we analyse how patient-initiated innovations contribute to

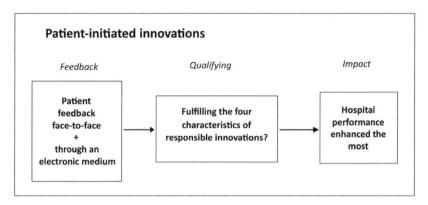

Figure 5.2   The relationship between patient feedback, responsible innovation and hospital performance

hospital performance in particular cases. We see six stages in the development of sustainable patient-initiated innovations:

1. Patients have no access to patient portals. Feedback is only given in face-to-face encounters.
2. Patients have access to their EHRs.
3. Feedback from patients increases.
4. The hospital utilizes patient feedback in a conscious effort to improve performance.
5. The hospital has a system to ensure that the follow-up of patient feedback fulfils the four characteristics of responsible innovation.
6. The hospital crowdsources ideas by issuing an open call for feedback to a group of current and/or former patients.

5.10   HOW TO STUDY PATIENT-INITIATED INNOVATIONS

One of the main purposes of this chapter is to present a research agenda for studying innovations in hospitals originating from patients' feedback. It is argued that there are trends pointing towards an increase in the number of feedbacks, and we argue that a proportion of these will be in the form of a proposal for change. The proposals from patients represent a potential for institutional change which may qualify as a patient-initiated innovation if it is shown to enhance hospital performance. We use the characteristics of 'responsible innovation' (Owen et al., 2013) as indicators of these kinds of innovations.

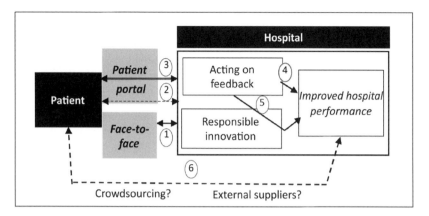

Figure 5.3   *Stages in the development of sustainable patient-initiated innovations*

We see three basic research questions in this area:

- What kind of relationship is there between the online release of health records to patients, and the design of feedback channels in a patient portal, and patient-initiated innovations?
- What impact do patient-initiated innovations have on hospital performance?

We focus on the impact of management strategies, and the design of the information systems for disseminating EHRs, and on procedures and systems for follow-up action after receiving feedback from patients.

We foresee that the patient feedbacks after accessing their EHR could be based on three types of reactions:

- A discrepancy between their perceived symptoms and their interpretation of the EHR.
- A discrepancy between their interpretation of the EHR and insights the patient believes they have acquired on his/her own.
- A discrepancy between the contents of the EHR and the patient's expectations, or requirements, with regard to the treatment procedure, or schedule.

We assume that many hospital patients engage when they are allowed to access their EHR. This assumption is supported by several studies on how patients respond when they are allowed to contribute to innovation processes in hospitals (Hibbard et al., 2004, Bullinger et al., 2012).

We should distinguish between two kinds of responses: Healthcare providers may respond *directly* or *indirectly* to all three categories of patient reactions. In direct responses healthcare providers act on information received directly (face-to-face or through an electronic medium) from a patient. In indirect responses healthcare providers act on information submitted by patients to an information system or to a person who files it into an information system. Healthcare providers then retrieve it from the information system.

At the receiving end, we focus on the hospital employees. We foresee four levels of response: The hospital employee:

- Listens to the patients, but does not take notes in any information system.
- Listens and take notes in an information system.
- Listens, take notes, and contacts colleagues to discuss the matter reported by the patient.
- Listens, take notes, and contacts colleagues with the intention of considering a change of hospital practice.

In addition, we foresee that the communication between the hospital and the patient can be initiated by the *hospital* in an effort to crowdsource[11] certain problems by issuing a general inquiry to a sample of patients.

These topics may be approached by applying two research designs:

- *Institutional analysis* of the patient-initiated innovation processes in hospitals.
  - Which patients (person background characteristics and diagnostic categories) are most active in this area?
  - What are the most effective models hospitals should implement in order to receive valuable feedbacks from patients?
- *Comparative studies* focusing on the significance of structures and policies that affect the rate, or content, of patient feedbacks, the transformation of patient feedback to patient-initiated innovations, and how such innovations may best enhance hospital performances. Here we study contextual differences between regions within and between countries.

5.11   SUMMING UP

Most of the research on how patients can contribute to innovations focus on patient–doctor communications, reducing costs, or empowering the individual patient to make the hospital treatment more effective. We find no major study on how patients may contribute to institutional innovations.

We highlight three basic research questions in this area and we foresee that patient feedback may be triggered by three types of patient reactions. In the follow-up at the receiving end in hospitals we distinguish between four levels of responses. We foresee six stages in the development of patient-initiated innovations. Patient-initiated innovations may be validated by considering the fulfilment of the four dimensions of 'responsible innovations' (Owen et al. 2013).

The relationship between patient-initiated innovations and hospital performance should be determined by considering how the patient-initiated innovations impact the hospital performance, but we cannot expect that these innovations always have a tangible impact on hospitals' formal performance variables. Empirical studies should be conducted using institutional analysis and/or a comparative design.

NOTES

1. This advice in the collection 'Decorum': "Perform all this calmly and adroitly, concealing most things from the patient while you are attending him" (Jones, 1923, p. 297).
2. When this text refers to 'patients', it is to be understood as the patient and any other close relative or friend supporting the patient.
3. The information that is disseminated to patients typically includes only certain parts of the 'complete' EHR. Personal notes, x-rays, and other elements that are considered inappropriate, are excluded. In this chapter 'EHR' denotes both the complete patient record, and shorter versions disseminated to patients.
4. 2015 was set as a time limit because it was not common to see fully implemented systems allowing patients' access to their EHR before this year.
5. The string used in Google Scholar was in the format: <keyword01> <keyword02> <keyword03> <keyword04> source: '<title Journal 01>' OR source: '<title Journal 02>' OR …
6. The search query returned 249 hits, but only 129 of these were from the journals specified in the search. This is because Google Scholar searches are based on keywords in the journal titles, not a database of actual journal titles. One journal in Table 5.1 includes additional articles due to similar keywords in the journal titles; this is 'Journal of Management'. This Is specified in the caption of Table 5.1.
7. The PAM instrument includes 22 items structured in four categories: 'believes active role is important', 'confidence and knowledge to take action', 'taking actions', and 'staying the course under stress'.
8. The interviews were conducted at the Stavanger Universitetssjukehus in the South-Western region of Norway. Interviews were semi-structured, based on an interview guide focusing on the hospital employees' experiences of patients giving feedback through electronic channels or in face-to-face contacts.
9. Schumpeter refers to five basic categories of innovation; the introduction of a new good, a new method for production, opening up a new market, the conquest of a new source of supplies, and carrying out a new organization (Schumpeter, 1983).
10. These eight principles are based on findings in the literature review in Hernandez et al. (2013): access, continuity, comprehensiveness, coordination, communication, cultural competency, family and person-focus, and payment alignment.
11. See Brabham (2008) for a discussion of how to delineate 'crowdsourcing' and its applications for problem solving.

REFERENCES

Ahmed, S., Ernst, P., Bartlett, S. J., Valois, M. F., Zaihra, T., Paré, G., Grad, R., Eilayyan, O., Perreault, R. & Tamblyn, R. (2016). The effectiveness of web-based asthma self-management system, My Asthma Portal (MAP): a pilot randomized controlled trial. *Journal of Medical Internet Research*, 18(12), online.

Augustine, S., Payne, B., Sencindiver, F. & Woodcock, S. (2005). Agile project management: steering from the edges. *Communications of the ACM*, 48(12), 85–89.

Austin, J. M., Jha, A. K., Romano, P. S., Singer, S. J., Vogus, T. J., Wachter, R. M. & Pronovost, P. J. (2015). National hospital ratings systems share few common scores and may generate confusion instead of clarity. *Health Affairs*, 34(3), 423–430.

Bangerter, B., Talwar, S., Arefi, R. & Stewart, K. (2014). Networks and devices for the 5G era. *IEEE Communications Magazine*, 52(2), 90–96.

Blackman, C. & Forge, S. (2016). European leadership in 5G. CEPS Special Report, December 2016. Report downloaded 27 April 2017 from http://www.europarl.europa.eu/RegData/etudes/IDAN/2016/595337/IPOL _IDA(2016)595337_EN.pdf.

Bos, L., Marsh, A., Carroll, D., Gupta, S. & Rees, M. (2008). Patient 2.0 Empowerment. *SWWS* (Semantic Web & Web Services), 164–168.

Brabham, D. C. (2008). Crowdsourcing as a model for problem solving: an introduction and cases. *Convergence*, 14(1), 75–90.

Bullinger, A. C., Rass, M., Adamczyk, S., Moeslein, K. M. & Sohn, S. (2012). Open innovation in health care: analysis of an open health platform. *Health Policy*, 105(2), 165–175.

Campanella, P., Lovato, E., Marone, C., Fallacara, L., Mancuso, A., Ricciardi, W. & Specchia, M. L. (2015). The impact of electronic health records on healthcare quality: a systematic review and meta-analysis. *European Journal of Public Health*, 26(1), 60–64.

Carinci, F., Van Gool, K., Mainz, J., Veillard, J., Pichora, E. C., Januel, J. M. … & Haelterman, M. (2015). Towards actionable international comparisons of health system performance: expert revision of the OECD framework and quality indicators. *International Journal for Quality in Health Care*, 27(2), 137–146.

Charles, C., Gafni, A. & Whelan, T. (1999). Decision-making in the physician–patient encounter: revisiting the shared treatment decision-making model. *Social Science & Medicine*, 49(5), 651–661.

Cohen, M. & Sundararajan, A. (2015). Self-regulation and innovation in the peer-to-peer sharing economy. *University of Chicago Law Review Dialogue*, 82, 116.

Conservative Party (UK) Manifesto 2017 (2017). Forward together, our plan for a stronger Britain and a prosperous future. Downloaded 19 May 2017 from https://order-order.com/wp-content/uploads/2017/05/Manifesto2017 .pdf.

Copnell, B., Hagger, V., Wilson, S. G., Evans, S. M., Sprivulis, P. C. & Cameron, P. A. (2009). Measuring the quality of hospital care: an inventory of indicators. *Internal Medicine Journal*, 39(6), 352–360.

Cleary, P. D. (2016). Evolving concepts of patient-centered care and the assessment of patient care experiences: optimism and opposition. *Journal of Health Politics, Policy and Law*, 41(4), 675–696.

Cucciniello, M., Guerrazzi, C., Nasi, G. & Ongaro, E. (2015). Coordination mechanisms for implementing complex innovations in the health care sector. *Public Management Review*, 17(7), 1040–1060.

Davies, H. T. & Lampel, J. (1998). Trust in performance indicators? *Quality in Health Care*, 7(3), 159–162.

Desouza, K. C., Awazu, Y., Jha, S., Dombrowski, C., Papagari, S., Baloh, P. & Kim, J. Y. (2008). Customer-driven innovation. *Research-Technology Management*, 51(3), 35–44.

Drejer, I. (2004). Identifying innovation in surveys of services: a Schumpeterian perspective. *Research Policy*, 33(3), 551–562.

Epstein, R. M., Campbell, T. L., Cohen-Cole, S. A., McWhinney, I. R. & Smilkstein, G. (1993). Perspectives on patient–doctor communication. *Journal of Family Practice*, 37, 377–377.

Eurostat (2018). Digital economy and society. Last updated 15 March 2018. Table 'individuals – internet use' (isoc_ci_ifp_iu). Downloaded 25 May 2018 from http://ec.europa.eu/eurostat/web/digital-economy-and-society/ data/database.

Garrido, T., Raymond, B. & Wheatley B. (2016). Lessons from more than a decade in patient portals. *HealthAffairsBlog*. Downloaded 21 September 2016 from http://healthaffairs.org/blog/2016/04/07/lessons-from-more-than -a-decade-in-patient-portals/.

Greene, J. & Hibbard, J. H. (2012). Why does patient activation matter? An examination of the relationships between patient activation and health-related outcomes. *Journal of General Internal Medicine*, 27(5), 520–526.

Hagel III, J. & Brown J. S. (2013). *Institutional innovation. Creating smarter organizations to scale learning*. Report published by Deloitte University Press. Downloaded 8 March 2019 from https://www2.deloitte.com/insights/ us/en/topics/innovation/institutional-innovation.html.

Hargrave, T. J. & Van de Ven, A. H. (2006). A collective action model of institutional innovation. *Academy of Management Review*, 31(4), 864–888.

HealthMine (2016). The 2016 HealthMine Digital Health Report. Downloaded 21 September 2016 from http://www.healthmine.com/research.

Hernandez, S. E., Conrad, D. A., Marcus-Smith, M. S., Reed, P. & Watts, C. (2013). Patient-centered innovation in health care organizations: a conceptual framework and case study application. *Health Care Management Review*, 38(2), 166–175.

Hibbard, J. H. & Greene, J. (2013). What the evidence shows about patient activation: better health outcomes and care experiences; fewer data on costs. *Health Affairs*, 32(2), 207–214.

Hibbard, J. H. & Mahoney, E. (2010). Toward a theory of patient and consumer activation. *Patient Education and Counselling*, 78(3), 377–381.

Hibbard, J. H., Stockard, J., Mahoney, E. R. & Tusler, M. (2004). Development of the Patient Activation Measure (PAM): conceptualizing and measuring activation in patients and consumers. *Health Services Research*, 39(4p1), 1005–1026.

Honeyman, A., Cox, B. & Fisher, B. (2005). Potential impacts of patient access to their electronic care records. *Journal of Innovation in Health Informatics*, 13(1), 55–60.

House of Commons (2016). A paperless NHS: electronic health care records. Briefing Paper Number 07572, 25 April 2016. Downloaded 26 October 2016 from http://researchbriefings.files.parliament.uk/documents/CBP -7572/CBP-7572.pdf.

Howrey, B. T., Thompson, B. L., Borkan, J., Kennedy, L. B., Hughes, L. S., Johnson, B. H. … & Degruy, F. (2015). Partnering with patients, families, and communities. *Family Medicine*, 47(8), 604–611.

Huvila, I., Myreteg, G. & Cajander, Å. (2013). Empowerment or anxiety? Research on deployment of online medical e-health services in Sweden. *Bulletin of the American Society for Information Science and Technology*, 39(5), 30–33.

Høj, P. (2009). Patienter kan se journaler på nettet. Downloaded 18 May 2016 from http://sn.dk/Sjaelland/Patienter-kan-se-journaler-paa-nettet/artikel/ 3284.

Jones, W. H. S. (1923). Hippocrates Volume II. The Loeb classical library. Downloaded 22 September 2016 from https://ryanfb.github.io/loebolus -data/L148.pdf.

Katz, J. (1984). *The Silent World of Doctor and Patient*. Maryland: The Free Press.

Klazinga, N., Fischer, C. & Ten Asbroek, A. (2011). Health services research related to performance indicators and benchmarking in Europe. *Journal of Health Services Research & Policy*, 16(2_suppl), 38–47.

Laudal, T., Bjaalid, G. & Mikkelsen, A. (2016). Hairy goals and organizational fit: a case of implementing ICT-supported task planning in a large hospital region. *Journal of Change Management*, 1–23.

Ledford, C. J., Cafferty, L. A. & Russell, T. C. (2015). The influence of health literacy and patient activation on patient information seeking and sharing. *Journal of Health Communication*, 20(sup2), 77–82.

Lindsay, C., Findlay, P., McQuarrie, J., Bennie, M., Corcoran, E. D. & Van Der Meer, R. (2018). Collaborative innovation, new technologies, and work redesign. *Public Administration Review*, 78(2), 251–260.

Lupton, D. & Jutel, A. (2015). 'It's like having a physician in your pocket!' A critical analysis of self-diagnosis smartphone apps. *Social Science & Medicine*, 133, 128–135.

National Health Service (NHS) Digital (2016). What is a Summary Care Record? Downloaded 26 October 2016 from http://systems.digital.nhs.uk/scr/patients/what.

Norwegian Directorate of eHealth (2016). One citizen, one record [presentation]. Downloaded 26 October 2016 from https://ehelse.no/Documents/E -helsekunnskap/One%20Citizen_One%20Record_Are_Muri.pdf.

Office of the National Coordinator for Health Information Technology (ONC) (2015). Adoption of electronic health record systems among US nonfederal acute care hospitals: 2008–2014. ONC Data Brief, No. 23. Downloaded 27 April 2017 from https://www.healthit.gov/sites/default/files/data-brief/2014HospitalAdoptionDataBrief.pdf.

Oram, A. (2001). *Peer-to-Peer: Harnessing the Power of Disruptive Technologies*. USA: O'Reilly Media, Inc.

Owen, R., Stilgoe, J., Macnaghten, P., Gorman, M., Fisher, E. & Guston, D. (2013). A framework for responsible innovation. In Owen, R., Bessant, J. and Heintz, M. (eds), *Responsible Innovation: Managing the Responsible Emergence of Science and Innovation in Society*. Chichester: John Wiley & Sons, 27–50.

Popov, S. (2016). The tangle. IOTA Whitepaper. Downloaded 16 June 2017 from https://iota.org/IOTA_Whitepaper.pdf.

Ricciardi, W. & Boccia, S. (2017). New challenges of public health: bringing the future of personalised healthcare into focus. *European Journal of Public Health*, 27(suppl_4), 36–39.

Robertson, A., Cresswell, K., Takian, A., Petrakaki, D., Crowe, S., Cornford, T. ... & Prescott, R. (2010). Implementation and adoption of nationwide electronic health records in secondary care in England: qualitative analysis of interim results from a prospective national evaluation. *British Medical Journal*, 341, c4564.

Schnipper, J., Gandhi, T., Wald, J., Grant, R., Poon, E., Volk, L. ... & Middleton, B. (2008). Design and implementation of a web-based patient portal linked to an electronic health record designed to improve medication safety: the Patient Gateway medications module. *Journal of Innovation in Health Informatics*, 16(2), 147–155.

Schumpeter, J. A. (1983), *The Theory of Economic Development*. New Jersey: Transaction Publishers. (Original version published in 1934.)

Simula, H. (2007). Concept of innovation revisited: a framework for a product innovation. *International Association for Management of Technology IAMOT 2007 Proceedings*.

Stewart, M. (2001). Towards a global definition of patient centred care: the patient should be the judge of patient centred care. *British Medical Journal*, 322(7284), 444.

Teodoro, R. R. (2016). *Beyond exposure: patient engagement with health information in an information ecology framework*. Doctoral dissertation, Rutgers University-Graduate School-New Brunswick.

Thomke, S. & von Hippel, E. (2002). Innovators. *Harvard Business Review*, 80(4), 74–81.

Topol, E. (2015). *The Patient Will See You Now: The Future of Medicine is in Your Hands*. New York: Basic Books.

Veldhuijzen, W., Elwyn, G., Ram, P. M., van der Weijden, T., van Leeuwen, Y. & van der Vleuten, C. P. M. (2011). Beyond patient centeredness: the multitude of competing goals that doctors pursue in consultations. Unpublished article included as Chapter 6 in Veldhuijzen, W., *Challenging the Patient Centred Paradigm: Designing Feasible Guidelines for Doctor Patient Communication*, Doctoral Dissertation, Maastricht University.

Vukovic, V., Parente, P., Campanella, P., Sulejmani, A., Ricciardi, W. & Specchia, M. L. (2017). Does public reporting influence quality, patient and provider's perspective, market share and disparities? A review. *European Journal of Public Health*, 27(6), 972–978.

Xu, X., Weber, I., Staples, M., Zhu, L., Bosch, J., Bass, L. … & Rimba, P. (2017). A taxonomy of blockchain-based systems for architecture design. In *Software Architecture (ICSA), 2017 IEEE International Conference* (pp. 243–252). IEEE.

6. University of Virginia health system's MyChart: supporting patient care and research

Bala Mulloth and Michael D. Williams

6.1 INTRODUCTION

Centralizing clinical data in standard, flexible, and accessible formats is the first step in harnessing the power of that data on behalf of clinicians and patients. UVA's initiatives around the MyChart portal are a progressive effort to drive this revolution in healthcare, by leveraging machine learning, algorithm, natural language processing and human observation to distil actionable insights from those data on behalf of our patients.

Jeffrey W. Keller, PhD, Chief Innovation Officer, University of Virginia Health System

The patient is the most underutilized healthcare resource (Garrido, Raymond and Wheatley, 2016), but recognition of the importance of patient engagement is increasing in health delivery. In this chapter we explore the following research questions: How can university health systems tap into benefits of health technologies in a responsible manner to better manage patients and improve their outcomes using patient-empowered tools? What is the opportunity for such innovation to occur with the general Responsible Research and Innovation (RRI) framework? With the ever-increasing amount of patient-generated data from connected medical devices, apps, and sensors, engaging patients via these devices seems like an obvious progression for the healthcare industry. Health information technologies such as Electronic Health Records (EHR), patient portals, patient engagement platforms, and remote monitoring have the greatest potential to empower and enable patients with their own clinical information and self-care for long term management. Patient portals can enhance patient-provider communication and enable patients to check test results, refill prescriptions, review their medical records, and view educational materials. Patient portals can also be used to streamline administrative tasks such as registration, appointment scheduling and patient reminders (Al Knawy, 2017).

It must be noted that healthcare systems are increasingly being evaluated on patient satisfaction and safety rather than by the quantity of patients seen. The high cost of medical malpractice suits to a healthcare system due to medical errors, which result in funds flowing out of a healthcare system entirely, as well as the patient overturn as a standard rating measure have sparked new conversations and research concerning how a healthcare system should operate efficiently and safely (McMahon et al., 2005). One inefficient and redundant factor in the healthcare system that limited timely communication between providers and amongst physicians, patients, and staff, as well as consuming costly real estate and physical medium through storing and printing is that of patient medical records (Poissant et al., 2005).

The Health Information and Technology for Economic and Clinical Health Act seeks to eliminate the redundant aspect of the health system and replace it with something more efficient and patient-centred. This Act became signed into law requiring every public and private healthcare institution in the US to use an Electronic Health Record (EHR) system to store, integrate, and consolidate patient protected health information. Incentives were provided to healthcare institutions to implement the required system into practice by 1 January 2015 or face negative adjustments for not doing so (Doran, Maurer and Ryan, 2017). Further, as part of the federal EHR Incentive Program meaningful use requirements were updated to improve patients' ability to view online, download, and transmit their health information via a patient portal. Physicians must also enable secure email exchange with patients.

The barriers that stand between the rapidly-developing commercial market for interoperable EHRs and improved patient outcomes, including their experience of care are considerable. Unfortunately, the second half of 2015 brought disappointing news regarding patient portal and secure email. A recent study (Garrido, Raymond and Wheatley, 2016), found that only 10.4 per cent of US hospitals met the current meaningful use objective of providing patients with online access to view, download, and transmit information about an admission. The same study also mentioned that a Nielsen Company survey of consumers found that only 15 per cent said they have access to email with their physician, and just over 20 per cent have access to online appointment scheduling.

In this chapter we aim to address the above-mentioned research questions using the example of an EHR implementation at the University of Virginia. The methodology employed is qualitative in nature and draws on evidence based on interpretative interviews as well as direct and indirect observations. The chapter is structured as follows. First, we provide an overview of EHR along with a use case. Subsequently we provide a description of the UVA MyChart system and ways it supports patient care and research. We then offer our analysis and lessons learned. Finally, we present the challenges to realize the full potential of such an EHR platform along with our concluding thoughts that exemplifies some of the unrealized opportunity for innovation in this space.

6.2 OVERVIEW OF ELECTRONIC HEALTH RECORDS (EHR)

EHR is viewed as the backbone supporting the integration of various information tools (e.g., emergency information, test ordering, electronic prescription, decision-support systems, digital imagery, and telemedicine) that could improve the uptake of data into clinical decisions (Gagnon et al., 2014). Using such evidence in daily clinical practice could enable a safer and more efficient healthcare system (Lam, Lee and Chen, 2016). International literature supports several benefits of EHR for patients (Samaan et al., 2009). One of the main benefits reported is the increased quality of care resulting from patients having their essential health data accessible to their different providers, which can significantly improve the coordination of care and increase the efficiency of primary care practice. There was also evidence that the EHR could support empowered citizens to actively take part in decisions regarding their health, and could be used to track the delivery of recommended preventive care across primary care practices (De Leon and Shih, 2011). The EHR is also a tool that facilitates knowledge exchange and decision making among healthcare professionals by providing them with relevant, timely, and up-to-date information (Canada Health Infoway, 2006).

In 2011, the Centers for Medicare & Medicaid Services (CMS), a branch of the Department of Health and Human Services established the Medicare and Medicaid Electronic Health Record (EHR) Incentive Programs to encourage Eligible Professionals, Eligible Hospitals, and Critical Access Hospitals to adopt, implement, upgrade, and demonstrate meaningful use of certified EHR technology (CEHRT). The EHR Incentive Programs consist of three stages (82 FR 87990):

Stage 1 set the foundation for the EHR Incentive Programs by establishing requirements for the electronic capture of clinical data, including providing patients with electronic copies of health information.

Stage 2 expanded upon the Stage 1 criteria with a focus on advancing clinical processes and ensuring that the meaningful use of EHRs supported the aims and priorities of the National Quality Strategy. Stage 2 criteria encouraged the use of CEHRT for continuous quality improvement at the point of care and the exchange of information in the most structured format possible.

In October 2015, CMS released a final rule[1] that modified Stage 2 to ease reporting requirements and align with other quality reporting programs. The final rule also established Stage 3 in 2017 and beyond, which focuses on using CEHRT to improve health outcomes.[2]

Physicians are likely to intend to use EHR when it is considered easy to use in their practice. Physicians know the importance and the necessity of using EHR in their practice, given its potential impact on the quality of care to patients (Buntin et al., 2011). Kaiser Permanente, one of the United States' largest not-for-profit health plans and an institution that has used portals for over a decade, reports that as of the third quarter of 2015, about 70 per cent (5.2 million patients) of eligible adult members registered to use its My Health Manager patient portal (Garrido, Raymond and Wheatley, 2016). It must be noted, however, that because EHR may lead to decreased productivity, EHR systems should be designed to take into account physician workload and cognitive capabilities (Boonstra and Broekhuis, 2010). Understanding which EHR features contribute to stress and burnout can help predict unintended consequences for physicians (Babbott et al., 2013) and offer opportunities to optimize EHR features that complement and enhance physician work life (Gagnon et al., 2014).

A special use case can be found in the EHR system in use by the Veteran's Administration (VA) which is charged with the care of US Servicemen and women. As is often the case in the US, advances in the military served as precursors for applications in the civilian world. The concept that eventually became one of, if not the first EHRs was initiated and planned at the beginning of the 1970s by the National Center for Health Services Research and Development of the US Public Health Service, now known as the Agency for Healthcare Research and Quality. Under the farsighted leadership of the VA's Chief Medical Director, Dr John Chase, the VA's Department of Medicine and Surgery (now known as the Veterans Health Administration (VHA)), then agreed to deploy the system at the largest medical system of that time, the VA hospitals. Today, the VA uses a platform known as My HealtheVet (https://www.myhealth.va.gov/). An excerpt from their website reads:

> Members registered in My HealtheVet are able to manage their health by using My HealtheVet's Blue Button feature to customize a report or access their VA Health Summary. The VA Blue Button feature helps you better manage your health care needs and communicate with your health care team.

6.3   MYCHART

With the adoption of Epic as their only EHR system beginning in 2011, the University of Virginia (UVA) medical centre's Epic integration process has undergone multiple evaluations with the purpose of optimizing their healthcare delivery system during the age of increasing Internet use and capabilities; an underlying infrastructure that did not exist or became too costly to implement in the past due to limitations of tech bandwidth, and provider barriers such as security concerns and hesitations for using new technology depending on when the physician's initial training occurred while receiving their degree. Epic's mantra is 'Epic, with the patient at the heart'. The University of Virginia Health System's (UVAHS's) unique patient population spanning across a wide geographic region served with many rural as well as urban communities, only heightened the need to conduct research around the implementation and practice of utilizing EHR's. The results and conclusions from past research focuses were centred primarily around expanding access to care by use of tele-medicine, increasing patient safety by reducing medical errors, and reducing costs by controlling for redundant systems that yielded healthcare expenditures so high that affordability for the patient became an unachievable milestone. In order to optimize the Epic experience with the patient at the heart, UVAHS gathered evaluations and assessments from past research, consisting of surveys and interviews from utilizing parties, and turned those results into a working model solely meant to enhance the patient experience. The working result consisted of utilizing a patient engagement capability of Epic known as MyChart.

MyChart gave patients the necessary tools required to be healthier and more involved throughout the patient/provider encounter. Putting records into patients' hands is not a new idea (Coleman, 1984). Baldry et al. (1986) conducted an early experiment in giving patients in the waiting room their medical records to read, and patient-held records seem to have ethical and practical benefits. MyChart's goal was to enhance communication with patients, decrease phone calls, reduce overlooked lab results, and improve documentation of patient history. The technological specifications for using the UVA MyChart capability of Epic requires that one only have access to a computer browser with cookies and scripting enabled. As long as the aforementioned requirement for using MyChart is met, patients and their family members will have gained access to an entirely new realm of healthcare that previously required costly specialized equipment, printing information on physical mediums, time-consuming communications spent between alternating worker shifts, and in-person patient consults/imaging requests that could take weeks to find if trying to recall records that were subsequently stored in warehouses off premises.

MyChart revolves around three aspects:[3]

- Patients have personal and family health information at their fingertips. Patients can message their doctors, attend e-visits, complete questionnaires, schedule appointments, and be more involved in managing their health.
- Patients in the hospital can use a tablet to stay in touch with their care team, review their schedule, access personalized patient education materials, and request help.

- Prospective patients can become new patients through easy online scheduling.

From UVAHS's website the MyChart description states "patients are offered personalized and secure on-line access to portions of their medical records, and also enables them to securely use the Internet to help manage and receive information about their health". With MyChart patients were able to communicate with their provider's office, see their test results, request prescription renewals, view their recent visit information and pay their bill/track insurance payments, among other capabilities. For security concerns, it must be noted that MyChart at UVA took care to ensure that health information was kept private and secure. Access to information was controlled through secure access codes, personal IDs, and passwords. Each person controlled their password, and the account could not be accessed without that password. MyChart used the latest encryption technology with no caching to automatically encrypt the session with MyChart. Unlike conventional email, all MyChart messaging was done while you were securely logged on to their website. Additional security measures consisted of providers choosing what information was accessible through MyChart as well as excluding highly sensitive records from ever being posted to MyChart as per Virginia Law.

The test results were available online in a timely manner after the patients visit. In general, labs that were collected during admissions would be released to MyChart upon discharge. Labs collected during clinic visits would be released to MyChart three days after the results had been reviewed by the provider. Imaging results, including those from Cardiology and Radiology, would be released seven days after the results had been reviewed by the provider. Pathology reports would be released to MyChart 14 days after the results had been reviewed by the provider. Also available was a tool to use if a result was expected to be available but had otherwise not been posted. The in-sight tool labelled Get Medical Advice allowed a direct message to be to sent to the corresponding clinic in cases such as this.

The importance of the MyChart capabilities, especially for the vast geographic region that UVA serves for a patient population, allowed for individuals to plan visits accordingly, saving time, money, and distance travelled for simple follow-up consults that would have otherwise created an increased burden on the hospital census and patient livelihoods. As an evaluation tool for the MyChart system to assess patient satisfaction, as of April 2018 (results pending), the UVAHS deployed a Consumer Assessment of Healthcare Providers and Systems (CAHPS®) tool, which is an Agency for Health Research and Quality (AHRQ) programme that began in 1995 to evaluate patient satisfaction with a wide variety of care-related experiences, including MyChart. The specific tool is known as the Census-Based Surveying™ methodology developed by the Press Ganey company. Included in the validated CAHPS Survey tool that every ambulatory patient seen at the UVAHS are questions specifically related to the use of MyChart. In order to more thoroughly evaluate patient satisfaction and utilization of MyChart, a successive independent samples design will be used to draw multiple random samples from the patient cadre population at one or more times over a six-month period.

6.4   ANALYSIS AND LESSONS LEARNED

It was well understood that innovation involved creating value from ideas, and successful innovation resulted in diffusion to scale. A number of disruptive trends (Hwang and Christensen, 2008) had accelerated the rise of new models of innovation in recent decades, including flexible IP policies, university–industry collaborations through incubators, micro entrepreneurs, venturing, distributed innovation in global locations, open innovation, and others. It was therefore important to consider the concept of responsibility in the context of innovation as a collective, uncertain, and future-oriented activity (Owen, Bessant and Heintz, 2013). We now refer to the concept of Responsible Research and Innovation (RRI) that was described as a "transparent, interactive process by which societal actors and innovators became mutually responsible to each other with a view on the ethical acceptability, sustainability, and societal desirability of the innovation process and its marketable products in order to allow a proper embedding of scientific and technological advances in our society" (von Schomberg, 2011). This is particularly important because UVAHS is undertaking a parallel innovation while continuing to deploy MyChart. This process emphasizes a new focus on the Social Determinants of Health (SDH) in which factors such as Income Levels, Food Insecurity, Barriers to Education, Environmental Stressors related to Climate Change and others are documented and patient-specific Care Plans are created with the social context of each patient as the framework. These Care Plans will be captured and most easily shared with not only the patient, but other health service providers through the use of Epic and the MyChart App as the patient-facing interface. New attention is being given to SDH by the US Department of Health and Human Services, which has authority to regulate care and administer reimbursement for health services for participants in Medicare and Medicaid.

The Centres for Medicare and Medicaid Services (CMS) recently released the CMS Core Health-Related Social Needs Screening Tool that provides a framework for healthcare systems and physician practices to understand their patient's Social Context; this is due to the considerable body of evidence that supports the view that Social Context accounts for as much as 60 per cent of an individual's health outcomes. Legislation is under consideration that would link reimbursement for Medicare-contracted services to SDH Risk Adjustment. As a result, the societal and economic imperatives for adopting an RRI framework that enables and emboldens health systems in collaborating with patients to proactively and innovatively evolve EHR platforms such as MyChart has taken on even greater urgency.

Utilizing the four core principles viz. anticipation, reflexivity, inclusivity and responsiveness of the RRI framework, we analyse below the UVAHS MyChart.

There had been a perceptible shift away from the single organization as a core source of technological innovation and a move towards using networks such as the UVAHS MyChart by leveraging its ecosystems to enhance innovation. This ecosystem emphasis by organizations such as the UVAHS and MyChart's focus on 'reflexivity' and 'inclusivity', for instance, was a growing trend and presented firms in an ever-increasing number of sectors with significant opportunities and challenges (Adner, 2006). It was increasingly clear that change would not occur when single organizations operated in a vacuum, and when they failed to understand the needs and perspectives of all relevant stakeholders. New networks of innovation and changing mindsets among people had a distinctive way of 'reperceiving' many of the enormous and urgent challenges in healthcare services into opportunities to "leverage the power of markets and business to have transformative, system-wide impacts" (Inslee and Hendricks, 2007, p. 21). Specifically, it had been noted that the UVAHS was adopting an innovative approach to achieving large-scale systematic change through 'collective impact' (Turner et al., 2012). Unlike traditional cross-sector collaboration, collective impact embodied a holistic approach to large-scale achievement and was underpinned by five key conditions for success: a common agenda; shared measurement; mutually reinforcing activities; continuous communication; and lastly, backbone support (Turner et al., 2012). Fostering collective impact not only required continuous communication and transparency among the various parts of the UVAHS but was also contingent upon the leadership provided by a central organization to serve as the 'backbone' for the entire initiative. Kaiser Permanente offered four key sessions from 10 years of experience with patient portals (Garrido, Raymond and Wheatley, 2016):

1. Secure email supports improved outcomes and patient-centred care.
2. Patient portal use positively impacts patient loyalty to the health plan and member satisfaction.
3. Evidence of a relationship between secure email and other kinds of utilization is mixed.

4. Even with the best intentions, e-health disparities can emerge. Members with the highest rate of registration and use of the portal are between 60 and 68 years old, white (non-Hispanic), educated to postgraduate level.

Based on our interviews with key stakeholders, including the Chief Innovation Officer and the Chief Medical Information Officer of the UVAHS, the UVAHS MyChart represented an initiative focused on collective impact and responsible innovation. By providing patients with personal and family health information at their fingertips in a safe and secure manner, MyChart strove to distribute key information to patients and partners in the medical and research communities quickly and efficiently. MyChart enabled multi-actor and public engagement in the healthcare processes, enabling easier access to medical records and instantaneous access to personalized patient education materials. In fact, MyChart even strove to encourage user-driven innovations. End users were considered important sources of clinically valuable innovations, both as healthcare professionals (Oliveira, von Hippel and DeMonaco, 2011) and patients and caregivers (Zejnilovic, Oliveira and Canhão, 2016). It was hypothesized that the 'next iteration of healthcare' would build on empowered patients – with their ability to innovate – to manage their own health and disease to ultimately encourage others to remain healthy or cope better with their health condition. MyChart sought to not only bring patients to the centre of decision making but also increase their responsibility and was, thus, to be viewed as social innovation.

In addition to providing stakeholders with information on the latest developments in the field, UVAHS through its MyChart served as a technological intermediary between doctors and patients as well as organizations and firms and had facilitated several technology-development initiatives totalling in investments worth several hundreds of dollars. In that sense, MyChart was closer to a new paradigm in public engagement with technology, a deepening of relationships and responsibilities of societal stakeholders for granting patients a 'license to participate'.

Further, it was also worthwhile to consider that, in many cases, technologies with a competitive advantage such as MyChart and a defined need failed, or they did not have the impact on the market that was predicted. Evolutionary theory of the firm (Hodgson,1998) argued that in many cases innovations found boundaries before their establishment, and the reason was not only that an innovation could be 'good or bad', but that there was a selection mechanism through a system of norms and rules written to 'trap or release' an innovation. Thus, even though innovations provided improved experience through better solutions, several promising innovations were never adopted at all, and others were subsequently abandoned.

Organizations such as the UVAHS played a critical role in anticipating such scenarios and by being 'responsive' in driving political, environmental, and social factors by involving different stakeholders with UVA and defining success criteria as per RRI (Owen, Bessant and Heintz, 2013). For example, as the MyChart technological capabilities evolved, the number of instances where it could empower patients became quite large. However, the UVAHS did not have the resources and expertise to focus on all of them. There were more barriers in how 'open' the system could be, as well as the time and training involved to get clinicians to learn a new 'craft', or the transfer of such process from one healthcare department to another. At this stage, the UVAHS had to make strategic decisions about where to focus its resources. It decided to focus primarily on enabling patients to message their doctors, attend e-visits, complete questionnaires, schedule appointments, and be more involved in managing their health. The thinking was that if the MyChart could demonstrate good user adoption for these functions, it would make getting clearance for other indications much easier in the future.

Also, UVAHS focused on championing the usage of the MyChart technology within and across a wide geographic region serving many rural as well as urban communities. The fact that the UVAHS had targeted remote areas in Virginia quite aggressively was a strong indication of its socially driven motivations. Further, the organization also focused on areas where there would be easy wins (i.e., communicate with your provider's office, see your test results, view your recent visit information, among other functions) where they were solely aiming to improve the patient's basic experience.

UVAHS also enabled responsible research as an innovation process where societal actors (researchers, citizens, policy makers, business, third-sector organizations, and so forth) worked together during the whole research and innovation process, both in negotiating which types of research and innovation to pursue and in subsequently (or simultaneously) developing research and innovation. In the healthcare sector, the federal government, through the National Institutes of Health, often shouldered major funding for initial research. Once the potential of new medical technologies was validated, corporations and venture capitalists were financially motivated to support critical steps towards commercialization. The transition between such basic research and actual technology application included the vital period of translational research.

The RRI concept acknowledged the power of research and innovation as a mechanism for genuine and transformative societal change to shape our collective future. In the words of Geoghegan-Quinn (2012), "Research and innovation must respond to the needs and ambitions of society, reflect its values and be responsible … Our duty as policy-makers [is] to shape a governance framework that encourages responsible research and innovation." It was here, in the translational phase, that researchers took risks that required resources (time, funding) to develop basic scientific concepts into revolutionary new clinical approaches for the prevention, diagnosis, and treatment of disease. The phase of translational research was traditionally underfunded, and was referred to as 'the Valley of Death'. The UVAHS had significant impact at this critical phase by not only anticipating and funding MyChart but by also being responsive and building a network of public and private partners both inside and outside of the medical industry equipped with the incentives, expertise, know-how, and resources required to successfully develop, commercialize, or apply technology relevant to MyChart developmental needs.

6.5 CHALLENGES AND CONCLUDING THOUGHTS

The single-largest contingency that would impact the future success of the MyChart was whether medical practitioners, government regulators, and consumers would adopt the technology as a viable solution for enhancing patient experience. As an established thought leader and intellectual hub within the space, the UVAHS was well positioned to leverage the knowledge and resources it had gained in order to finance the continued innovation of MyChart technology within the medical field. It had certainly made an impact – approximately 3 per cent of nearly 1 million outpatient visits are associated with use of MyChart – the UVAHS had largely contributed to the cause through trial and experimentation.

However, in order to craft a more sustainable user adoption model, the UVAHS could have had to consider pursuing a method to more thoroughly evaluate patient satisfaction and utilization of MyChart. In order to carry out this strategy, a successive independent samples design will be used to draw multiple random samples from the patient care population at one or more times over a six-month period. As with many organizational leaders of collective impact, the role of the UVAHS would surely shift over time. Once an overarching vision and strategy had been established within the MyChart community, the technology would likely pivot from delivering basic patient experience improvement to the advancement of digitization – this was particularly relevant within the healthcare field. As meaningful use continues to evolve, and as consumer expectations for user-friendly technology increase, the next few years will be critical for patient portals such as MyChart. We believe that patient empowerment tools help our healthcare system improve outcomes and manage resources.

Considerable future challenges remain in order to realize the full potential the EHR platform represents. Some of the most pressing include e-billing. Consumers of healthcare services are shouldering an increasing share of the cost of care as employer-sponsored third-party insurance products offer fewer benefits at higher premiums from employers. At the same time, disruptive platforms such as Uber and Lyft are allowing consumers to bypass the proverbial middle-man and e-pay in advance of the delivery of the ride service, based on a pre-negotiated price. MyChart now offers a bill pay feature with e-reminders delivered through the smartphone app which allows customers to pay their proportion of the cost of health services electronically using a stored bank account number or credit card. As price transparency becomes an increasing demand of consumers, US health systems, including UVA, will need to leverage EHRs to not only collect payment, but explain/justify them. This challenge may be unique to commercially-driven systems like those in the US.

Additionally, physician adoption at the individual and small/private practice level has been difficult as the negative financial impact on a small practice has delayed/prevented implementation of EHRs, as mentioned above. As large non-health sector business interests increasingly turn their attention to this space (e.g., Amazon, Walmart, JP Morgan and others) the market forces and therefore focus may shift away from one-on-one doctor/patient relationships and drive the need for even greater personalization and portability of individual medical records. The ability for consumers to own all of their medical information and share it on their terms with any provider at any time becomes of paramount importance. Current levels of interoperability render this impossible. The opportunity for innovation within an RRI framework is considerable and timely.

Lastly, as technology evolves at breakneck pace, the platform of electronic health records will too. Phone-based technologies, wearables, Wi-Fi enabled devices, all of these are likely (or currently) suitable infrastructure for the acquisition, communication and storage of health information. Securing this information from data/identity theft will be increasingly important to patients and vendors of EHRs alike. Widely-reported breaches in privacy at health systems and hospitals clearly demonstrate the need for another innovation in this space: how to secure *and still be able to share* sensitive health information? As firms wrestle with the technological challenges associated with this imperative, broader questions as to how different cultures value security will be faced. The successful innovator will necessarily design a product that can adapt to the various societal norms in which it will be deployed. Where individual privacy is held sacrosanct, a responsible product may compromise inter-facility data exchange. Where social norms render individual privacy subservient to the common good, a more flexible approach will be necessary. These questions will be intensely local and utilizing an RRI framework will be crucial.

NOTES

1. https://www.federalregister.gov/documents/2015/10/16/2015-25595/medicare -and-medicaid-programs-electronic-health-record-incentive-program-stage-3-and -modifications.
2. https://www.cms.gov/Regulations-and-Guidance/Legislation/EHR IncentivePrograms/index.html?redirect=/EHRIncentivePrograms.
3. Based on interviews conducted with key stakeholders, including the Chief Innovation Officer and the Chief Medical Information Officer of the UVAHS. Five interviews were conducted in the form of face-to-face meetings as well as phone conversations during the March 2017 to June 2018 time period.

REFERENCES

Adner, R. (2006). Match your innovation strategy to your innovation ecosystem. *Harvard Business Review*, 84(4), 98.

Al Knawy, B. A. (ed.) (2017). *Leading Reliable Healthcare*. London: Taylor & Francis.

Coleman, V. (1984). Why patients should keep their own records. *Journal of Medical Ethics*, 10(1), 27–28.

Babbott, S., Manwell, L. B., Brown, R., Montague, E., Williams, E., Schwartz, M. ... & Linzer, M. (2013). Electronic medical records and physician stress in primary care: results from the MEMO Study. *Journal of the American Medical Informatics Association*, 21(e1), e100–e106.

Baldry, M., Cheal, C., Fisher, B., Gillett, M. & Huet, V. (1986). Giving patients their own records in general practice: experience of patients and staff. *British Medical Journal (Clinical Research Edition)*, 292(6520), 596–598.

Boonstra, A. & Broekhuis, M. (2010). Barriers to the acceptance of electronic medical records by physicians from systematic review to taxonomy and interventions. *BMC Health Services Research*, 10(1), 231.

Buntin, M. B., Burke, M. F., Hoaglin, M. C. & Blumenthal, D. (2011). The benefits of health information technology: a review of the recent literature shows predominantly positive results. *Health Affairs*, 30(3), 464–471.

Canada Health Infoway (2006). Beyond good intentions: accelerating the electronic health record in Canada. In Canada Health Infoway and Health Council of Canada (CHIHCC) (ed.), Montebello QC. Retrieved on 8 March 2019 from https://healthcouncilcanada.ca/files/2.15-infoway.pdf.

De Leon, S. F. & Shih, S. C. (2011). Tracking the delivery of prevention-oriented care among primary care providers who have adopted electronic health records. *Journal of the American Medical Informatics Association*, 18(Supplement_1), i91–i95.

Doran, T., Maurer, K. A. & Ryan, A. M. (2017). Impact of provider incentives on quality and value of health care. *Annual Review of Public Health*, 38, 449–465.

Gagnon, M. P., Ghandour, E. K., Talla, P. K., Simonyan, D., Godin, G., Labrecque, M. ... & Rousseau, M. (2014). Electronic health record acceptance by physicians: testing an integrated theoretical model. *Journal of Biomedical Informatics*, 48, 17–27.

Garrido, T., Raymond, B. & Wheatley, B. (2016). Lessons from more than a decade in patient portals. *Health Affairs Blog*, 2016.

Geoghegan-Quinn, M. (2012). 'Winning the Innovation Race', lecture at 2012 Innovation Summit in Brussels, Belgium. Retrieved on 27 January 2018 from http://www.moublog.com/2012/12/winning-innovation-race-maire -geoghegan-quinn-eu-commissioner.html#sthash.9mxmSlCS.dpbs.

Hodgson, G. M. (1998). The approach of institutional economics. *Journal of Economic Literature*, 36(1), 166–192.

Hwang, J. & Christensen, C. M. (2008). Disruptive innovation in health care delivery: a framework for business-model innovation. *Health Affairs*, 27(5), 1329–1335.

Inslee, J. & Hendricks, B (2007). *Apollo's Fire: Igniting America's Clean Energy Economy*. Washington, DC: Island Press.

Lam, J. G., Lee, B. S. & Chen, P. P. (2016). The effect of electronic health records adoption on patient visit volume at an academic ophthalmology department. *BMC Health Services Research*, 16(1), 7.

McMahon, K., McCarthy, J., Goodson, R., Stein, J., O'Donnell, D. & Zyck, K. (2005). Health care technology's impact on medical malpractice. In *Crittenden Medical Insurance Conference*, May.

Oliveira, P., von Hippel, E. & DeMonaco, H. (2011). *Patients as Healthcare Innovators: The Case of Cystic Fibrosis*. Working Paper, MIT.

Owen, R., Bessant, J. & Heintz, M. (eds) (2013). *Responsible Innovation: Managing the Responsible Emergence of Science and Innovation in Society*. Chichester: John Wiley & Sons.

Poissant, L., Pereira, J., Tamblyn, R. & Kawasumi, Y. (2005). The impact of electronic health records on time efficiency of physicians and nurses: a systematic review. *Journal of the American Medical Informatics Association*, 12(5), 505–516.

Samaan, Z. M., Klein, M. D., Mansour, M. E. & DeWitt, T. G. (2009). The impact of the electronic health record on an academic pediatric primary care center. *The Journal of Ambulatory Care Management*, 32(3), 180–187.

Turner, S., Merchant, K., Kania, J. & Martin, E. (2012). Understanding the value of backbone organizations in collective impact: part 2. *Stanford Social Innovation Review*. Retrieved in May 2018 from https://ssir.org/articles/

entry/understanding_the_value_of_backbone_organizations_in_collective _impact_2.

von Schomberg, R. (2011). Towards responsible research and innovation in the information and communication technologies and security technologies fields. SSRN Electronic Journal. Retrieved in May 2018 from https://www .researchgate.net/publication/239917899_Towards_Responsible_Research _and_Innovation_in_the_Information_and_Communication_Technologies _and_Security_Technologies_Field.

Zejnilovic, L., Oliveira, P. & Canhão, H. (2016). Innovations by and for patients, and their place in the future health care system. In Horst Albach, Heribert Meffert, Andreas Pinkwart, Ralf Reichwald and Wilfried von Eiff (eds), *Boundaryless Hospital* (pp. 341–357). Berlin, Heidelberg: Springer.

7. Design space in digital healthcare – the case of health information TV

John Bessant, Allen Alexander, Danielle Wynne and Anna Trifilova

7.1 THE PROMISE – AND THE PROBLEM – IN DIGITAL HEALTHCARE

We saw earlier the challenge and opportunity facing healthcare. Whilst there are significantly increasing pressures on spending in order to try and maintain levels of delivery and improve services, these are countered by challenges posed by ageing populations, rising prices and increasing complexity of healthcare technology. Despite wide variations in the healthcare funding system, from largely public (as in the UK) to strongly private, the underlying trends are the same.

Against this backdrop the need for radical innovation is clear and extensive efforts are being made to find a way out of the crisis through new approaches. One powerful candidate within this field is the application of information and communication technologies on a wide scale – what we have been calling 'digital healthcare'.

Over the past twenty years these technologies have matured and converged to the point that there is now an explosion of innovative application. Their potential is significant – not just in terms of improving productivity within the healthcare delivery sector but also in offering better outcomes, higher quality and reliability, greater patient autonomy and higher quality of life. Taken at face value, digital healthcare appears to offer a rosy future for patient-centred high-quality healthcare delivery at an affordable cost and open to all.

7.1.1 Design Space and Technological Determinism

One of the problems under conditions like these is that innovations quickly converge into what is termed a 'dominant design' (Abernathy and Utterback, 1978; Utterback, 1994). Such convergence represents the resolution of a number of shaping forces but evidence suggests that dominant designs may not always be the 'best' or most optimal solution but rather the ones which emerge from a process of contestation around design. Dominant designs are not neutral – they encapsulate the relative power and strategic intent of different interest groups. There is a risk that alternative designs might be excluded at an early stage as a trajectory emerges, which defines the form and implementation mode of the technology – the challenge of 'technological determinism' (Braverman, 1974).

There are parallels to this in a number of other fields. Early adoption of computer-aided production management systems in the 1970s and 1980s involved an embedded model of how organizations worked and were structured, which suited certain kinds of application but limited the effectiveness of the technology in other contexts (Bessant and Buckingham, 1993). Similarly, flexible automation technologies during the 1990s often failed to deliver their potential because of the inbuilt design logic, which assumed certain forms of work organization (Bessant et al., 1992). Indeed the emergence of 'lean manufacturing;' with its emphasis on team working within flexible and autonomous teams owed much to its ability to deliver the flexibility which expensive but rigidly designed technologies could not (Womack and Jones, 1996).

Experiences like these suggest that there is an initial 'design space' associated with novel technological opportunities but that this can quickly become colonized by a dominant design and force out other options (Buchanan and Bessant, 1985). A counter-strategy is to engage in extensive engagement with stakeholders who will operate or be affected by these technologies at an early stage (Mumford, 1979; Eason, 1988; Leonard-Barton, 1988).

There is increasing evidence of the value which taking an inclusive user perspective can bring to innovation. In particular, work by von Hippel and colleagues has shown over decades that working with such 'user innovators' helps articulate 'sticky information' which might not otherwise be available in researching market potential and which can be used to enhance designs to make them more acceptable to a wider population (von Hippel, 2005; Harhoff and Lakhani, 2016; von Hippel, 2016). Addressing the 'compatibility' question is central to much of the theory of adoption and diffusion of innovation and working with lead users offers an important channel for so doing (Moore, 1999; Rogers, 2003).

Although there is an emerging rhetoric around patient engagement in the innovation process and recognition of the need to consider this perspective, there are significant institutional barriers to building patient voices into the design of medical innovation (Bate and Robert 2006; Bevan et al., 2007). In particular the logic of procurement in many healthcare systems is one of centralization and scale. Although programmable and flexible in theory, the nature of this logic argues for one size fits many kinds of solution. Pilots are then rolled out without subsequent tailoring or configuration to suit differing local circumstances (Crisp, 2010).

A third point of relevance in the healthcare field concerns *who* is undertaking the design activity. Unlike consumer markets where the interests of the user are important input to early design since this will shape downstream adoption and diffusion, in the medical field there is a two-step model in operation. Ideas are often initially developed with the concerns of clinicians in mind who are assumed to act on behalf of the end-recipient of care – the patient. The risk here is that one group of users is consulted but another is disenfranchised from participating in the design; the result can be a 'clinicians know best' solution, which may not meet the underlying patients' needs or concerns.

So whilst it is possible to envisage a utopian world of great healthcare for all at affordable prices, there is at least the possibility that an alternative dystopian view night also emerge. Examples might include 'smart' homes, which limit the autonomy of residents – effectively becoming high-tech prisons. Or privacy issues associated with misuse of electronic medical records. Or decision-making about access to healthcare being linked to algorithms within machine learning systems and disenfranchising access to care. Given this risk we suggest that digital healthcare is one of many technologies around which concerns of responsible innovation could be raised.

7.2   RESPONSIBLE INNOVATION

As we saw in Chapter 2, concern about the implications of technological decisions and the identification of mechanisms through which such decisions could be modulated by consideration of alternative outcomes and engagement of multiple stakeholders is not new. Its most recent manifestation can be seen in work on 'responsible innovation' (Owen, Bessant and Heintz, 2013).

Importantly the (RI) debate is not one characterized by an anti-technology or Luddite bias but rather one which focuses on asking key questions shaping those developments and applications at an early enough stage to influence the design and implementation. In other words it seeks to explore technological options at an early enough stage when the 'design space' is broad.

In this chapter we will focus particularly on the ways in which different perspectives and insights from different stakeholders can be brought into the innovation design space. To do so it will be helpful to explore three related strands of literature which offer perspectives on how such an inclusive approach might be implemented.

7.3   INCLUSIVENESS AS A CORE DIMENSION

As we saw above, technological systems do not simply appear – they emerge as the result of a social and technological shaping process involving multiple actors. These represent embedded ideas, values, perspectives – and by their nature exclude others. Innovation adoption and diffusion is partly linked to finding a good fit between the needs/expectations of the receiver population and the offer represented by the emerging dominant design.

There are many examples in history which highlight that dominant designs may not represent the interests of all stakeholders. Perhaps the most powerful image is that of Charlie Chaplin in the film *Modern Times* in which he plays a character who becomes literally a cog in a machine. This exaggerated image of the factory model typified by Henry Ford's mass production system suggests that whilst it was a highly effective, indeed transformative, approach to factory organization it did not offer the best in quality of working life!

Research-based studies have repeatedly shown this effect – the implementation of dominant designs which are effective but which may marginalize the interests and concerns of some stakeholders. Resistance to change is an inevitable consequence of this and much of the literature around socio-technical systems design and the later work on team-based automation and working tried to explore alternative models (Trist and Bamforth, 1951; Eason, 1988).

This intersects with a second contributing literature stream dealing with user involvement in the innovation process. Medical innovations have been widely studied within this context (von Hippel, 1988; Herstatt and von Hippel, 1992). As a growing range of studies have shown, users can represent an important force in shaping a wide range of product and process innovations; emphasis has now shifted to understanding the segmentation across user populations in terms of their interest in participating and in the toolkits available to enable wider user innovation (von Hippel, 2016). Within this context a number of studies explore the ways in which patients (and their carers) as 'end-users' can play a significant role in developing and modulating innovations (Bate and Robert, 2006; Bessant and Maher, 2009; Pickles, Hide and Maher, 2008; Habicht, Oliveira and Shcherbatiuk, 2013). Once again we try to link this to 'inclusiveness' within responsible innovation.

Mobilizing user insights in practice is not without its difficulties. Bringing multiple stakeholder views to the surface and exploring their concerns and insights takes time and may slow the design phase, setting up tension between time and resource pressure and the desire to open up the process.

A second challenge relates to the willingness and/or the ability of users to participate. Not every user wishes to be an innovator and there is a distribution across a spectrum of potential involvement. Linked to this is the need for tools and methods to enable user insights to be articulated and captured. Our discussion in Chapter 3 and elsewhere highlights the different roles which patients as 'end-users' might take, ranging form purely passive, through informed, actively involved right up to the kind of 'hero innovation' described in Chapter 4.

Nonetheless there is little doubt that bringing in user perspectives and insights early in the process can enhance the quality of design and its downstream acceptability. Literature explores this theme in a number of settings – for example in user-led innovation/free innovation, via early involvement of functions in product development or in the participative design of technological systems. (Bødker, Kensing and Simonsen, 2004; Baskerville and Myers, 2009). Mobilizing user insights in systematic fashion is a core element in 'design thinking' and much of the literature around 'change management' builds in similar principles. (Brown 2008; Bessant, Abu El-Ella and Pinkwart, 2015).

Further impetus is given to this approach by the major advantage of digital technologies which is their malleability; software can be adapted and configured to accommodate different perspectives within a contested design space (Tidd and Bessant, 2018, forthcoming). They have the potential to change the balance between 'richness' and 'reach', effectively opening up the innovation process to more players who can play an active role in shaping the outcomes (Evans and Wurster, 2000).

As discussions of 'open innovation' mature so it is becoming clear that there are modes of 'crowdsourcing' which can contribute both at the front (idea generation) end of innovation and also in downstream modification and diffusion (Brabham et al., 2014; Harhoff and Lakhani, 2016). This opens scope for 'democratization of innovation' within the field (von Hippel, 2005) and specifically for the engagement of more and different stakeholder perspectives – the 'inclusiveness' dimension in the Stilgoe, Owen and Macnaghten (2013) framework for responsible innovation.

The design process associated with digital technologies also allows for 'delaying the freeze' – extending the time and space available for exploring different prototypes and pivoting towards an acceptable solution which meets the needs of diverse stakeholders. This is a core feature in discussions of 'lean start-up' and 'agile innovation' (Ries, 2011; Blank, 2013; Morris, Ma and Wu, 2014.

7.4   RESEARCH DESIGN

Our case study involves the development and diffusion (with subsequent modification and 're-innovation') of a digital health information platform. This platform uses a wide range of short video and web-enabled information tools across which users and providers can share information and improve awareness of key approaches, communicate across relevant communities and in other ways enhance the availability and application of health education. The platform is growing rapidly and has widespread support across users and healthcare agencies; as such it represents a good case in which to explore the extent to which responsible innovation principles are or could be included.

The research is based on a series of interviews with the founders and key stakeholders involved in the early development (15 interviews) and with users and others who have shaped its subsequent development and diffusion (12 interviews). The case is supported by archive and other relevant documentary data. We began interviewing four years ago and have been tracking the evolution of both the innovation and the ways in which different stakeholders have shaped and modified it; this process is continuing, providing a useful longitudinal perspective.

7.5   THE STORY

The origins of the innovation in question go back to 2012 and its subsequent development can be mapped across three phases:

7.5.1   Early Stage

In late 2012 Peter Smith was motivated to try and develop improved access to medical information. Working within the healthcare system he was aware of the extensive specialized knowledge available but was concerned that its accessibility was often limited, needing expertise in both finding and understanding it. He had a particular interest since his father was a diabetic and Peter wanted him to understand more about his condition and how he could manage it. He conceived of the idea of HealthTV as a way of opening up access to such information. The core value proposition was based on providing reliable health information to those who needed it and in a form which they could use.

Early work with a local college helped explore the possibility of producing videos to meet these needs but it quickly became apparent after a couple of pilots that this route would prove too expensive. At a networking event Peter met with James Wilson who had recently founded a company specializing in short high impact videos – MiniTV. There seemed to be scope for exploring collaboration. Their discussions led to a revised concept – Librio – which by January 2013 took shape, based on providing short video-based information in three forms:

• General public health and wellbeing education;

- Providing specific 'pathway' knowledge, for example to parents learning how to look after a sick child; and
- Patient-focused information about their treatment journey through the NHS.

Interest grew in the concept within the healthcare delivery Trust in which Peter worked. A number of individuals and groups became involved, bringing experience and perspectives, including several user groups based within the education directorate for the entire Trust (covering hospitals, primary and community care, etc.). Several avenues were explored including e-learning, podcasts, online and mobile device formats and so on. Involvement of this kind brought with it a variety of experience about different delivery channels and media and the revised business model could be explored and tested.

The concept was elaborated during a variety of design sessions where it was challenged and improved. External events also provided a focus; a major review of healthcare in 2013 (the Francis Enquiry) argued strongly for improved health information and education and Librio played to the emerging strategic agenda for healthcare providers in responding to this. And on the technology side increasing growth in the idea of live streaming TV led to a derivative of Librio – Livestream – coming out in March 2013, using (and providing learning opportunities around) web streaming.

Funding for these development initiatives had been coming in an ad hoc fashion from the healthcare Trust with a significant investment of time and personal resource from Peter. In particular the Trust provided protected time for Peter and additional secondments to his team of 2 people for 0.4 of their time. It also provided funding for three activities developing a prototype (the Librio Public Health website), developing a test site for Librio Educate, and commissioning content to be created jointly in partnership with MiniTV.

Between October and December 2013 various versions of a bid for formal Business Development support were put together to try to secure a stronger funding base. The first part of the proposal highlighted progress made so far since Librio was launched in April 2013, with over 70,000 visitors to the website. This level of traffic enabled a robust test of the Librio Educate offering with potential customers and pilot activities (such as the 'Preparing Children for Paediatric Daycase Surgery') resulted in demonstrable cost savings for the Trust through reduced surgery cancellations.

The proposed next step was based on a streamlined business model with three core target 'businesses':

- Librio Public Health – drawing on a library of video, and other learning resources and delivered via multiple channels.
- Librio Educate – as above but more specifically targeted to key groups.

- Librio Digital Services – including video production, website development, app development, etc. Effectively making Librio's experience available to others on a contract basis.

In February 2014 this was presented to the Board of the Regional Health Authority.

7.5.2 Phase Two – Emerging Split

Although recognizing the significant progress which had been made the Board also had access to a strategic review which had been commissioned and which highlighted a number of concerns which would need to be addressed in taking the business forward. In particular these included:

- Lack of focus in the core strategy, trying to hit too many targets simultaneously;
- Scale of the challenge at national level and concern about resourcing to meet this; and
- Strong competition in the existing education market.

The overall conclusion was to push for a more targeted and focused strategy; after exploring several options for this focus including a narrower version of the public health offering and the use of tailored training packages within Librio Educate the main recommendation was a clear concentration on the Librio Pathway model. The argument behind this was based on the relative lack of competition in this space and the potential of a targeted set of products; importantly the concept of 'pathway' was redefined to have a narrower meaning linked to the clinical pathway rather than the wider patient experience pathway.

The discussion which followed this recommendation highlighted a number of strategic tensions amongst those involved. For Peter there was a growing tension between his vision – of providing wide-ranging access to health information – and that emerging around a tightly focused business proposition targeted principally at clinicians and people working within the health sector. For the Trust there was concern that the current model for Librio lacked clear strategy or focus and risked becoming a drain on resources with impact which was too diffuse.

This tension became a fault-line along which two differing models of how the business could develop began to emerge. An indication of this can be gained from analysing the content of the discussion and the issues raised – for example:

- Peter wanted to move to an inclusive TV environment in which people could discuss their condition with links enabling them to access specific pathway resources. Concerns here were around the scale of doing this and the use of the term 'pathway' to describe the whole patient experience rather than a narrower clinical definition.
- Tension around clarity of the business model, especially on whether Librio was offering a clear product or a range of services – part of the proposed new business plan involved offering a range of contract video and learning resource production services to others.
- Developing the business further would require an increased scale of investment, especially in areas like IP, commercial support, marketing and website development support. Spread across the whole range of activities proposed in the new business plan this would involve a significant investment so the preference was for a smaller more focused input.
- One of the options in moving to a business model focused tightly on pathways was the possibility of using video to help recruit to and manage clinical trials. This offered significant possibilities in terms of recruitment but also in terms of creating an income stream for the product. However, moves in this focused and commercial direction ran up against Peter's vision of a more open and inclusive model of disseminating health information.

The outcome of the meeting was a revised strategy involving several key decisions shaping the future of the business:

- **Live-Stream** – continue under review since the majority of the financial investment required for the live TV has already been made.
- **Pathway Video** – exploring the concept of bespoke clinical trials films as a separate, potentially licensable innovation.
- **Librio** – focus on patient Pathway films aimed at Prepare, Support and Recover.
- **MiniTV** – priority should be given to an assessment of whether or not a partnership between the Trust and MiniTV would be possible based on a formal contract.

During the following months these themes were explored further but it became increasingly clear that there were now two different business models involved. From Peter's side there was a change in direction and focus around Librio as a web-based TV provider of health information services. He wanted to develop this further and curate the material available, both in terms of the library developed so far and future offerings. From the Trust side the preferred model was pathway-focused and managed in a more formal way; there was also a potential brand management problem in continuing to offer two different approaches under the same label.

7.5.3 Phase Three – Divergence

By October 2014 the decision to support the pathway model had been made but Peter's increasing discomfort with this solution led to his deciding to split off Librio and his health information model from the programme. Negotiations with Paul around intellectual property and so on led to him being paid off and Librio continued to develop as an independent venture. Internally the Trust moved to create a new partnership with MiniTV to offer Health & Care videos for patient pathways – Southwest Health Video services. They developed a revised business model and tested the concept on clinicians during November 2014; responses were favourable and during 2015 the new business developed via some pilot projects using internal funding from the Trust and MiniTV development resources. By July 2016 the model was formally signed off by the senior management of the Trust.

Early operating experience highlighted a number of development points, especially around clinician support for learning how to use the resource effectively. The revised business model, including a clear strategic plan, pricing structure, marketing and technical development strategy, and so on involved:

- Targeting five key business lines:
  - Secondary care (e.g., hospitals)
  - Support to general practitioners
  - Clinical trials recruitment
  - Support to social care staff
  - Support for obtaining consent from people for procedures (including preparing patients for discussion with physician and letting people know about risks)
- Doubling turnover from the 2016 base
- Extending the reach by aiming for 40 per cent of commissioning to come from other healthcare Trusts
- Operating a for-profit business model but with a high degree of social consciousness.

Current progress of SDHV has followed this model, but with some revisions. Sales have come more from new products than from 'rebadged' stock in the old Librio library. The market has expanded to include several major Trusts in the UK, other actors in the healthcare space including publishers, pharmaceutical companies and healthcare platform providers) and, importantly, other sector agencies with a similar education/information challenge. There is still an issue around adoption, especially in getting clinicians and practitioners to use the videos easily within daily practice.

Whilst initially the proposition emphasized a mix of benefits, it has become more targeted on improving care pathways from an operational sense, with a strong focus on the cost-saving potential. Better-informed patients are more able to care for themselves, avoiding the need to interact with services, or when they do need to interact the duration is shortened. This generates capacity and delivers cost avoidance.

7.6   DISCUSSION

Responsible innovation is not a new concept; it has its roots in several traditions which are essentially concerned with careful review of the implications of particular innovations at an early enough stage to allow for some degree of shaping their development. The model offered by Stilgoe, Owen and Macnaghten (2013) suggests four key fields around which such discussions might take place and mapping our case study against it highlights a number of points:

- Anticipation – the process of developing a new venture requires imagining different futures. The power of a structured approach (in our case this was the Business Model Canvas (BMC) (Osterwalder and Pigneur, 2010)) is that it brings key questions into this discussion and provides a relevant anticipatory overview. However, even though questions may be raised, the ways in which they are answered may still reflect underlying beliefs or motives which are not open to challenge, effectively introducing an element of cognitive dissonance to the exploration. So it is important in using such frameworks to have diversity and perhaps even an explicit devil's advocate role, in order to challenge rigorously.
- Reflection – the underlying values held by the players and their organizations took time to emerge and were in conflict at key decision points, leading eventually to the splitting. Once again this argues for a structured review process which brings in and articulates different positions; essentially there emerged two competing business logics, one socially driven and one practically driven.
- Deliberation and inclusiveness – essentially the core value proposition began to change as the question of 'value for whom' was explored empirically. 'Value' moved from being a theoretical construct to how the project would continue to be funded – and this forced the issue around who to include or exclude in the design. Once again the use of a framework and

ancillary questions becomes a valuable device for exploring this and checking that different perspectives have been brought to bear. But again there is a need for facilitation and support for doing so in a robust fashion.

- Responsiveness – as suggested above, the process which any start-up has to go through offers a number of decision points where these issues can be considered. Structured frameworks offer a powerful enabling device for developing responsiveness but the underlying political and personal drivers remain. This means that there will always be a 'contested' aspect to exploring the 'design space' – and again argues strongly for careful and experienced facilitation of the process.

Viewed as a case history of a start-up we can see a number of interesting features. First is that there was no fully developed 'business model' for the original idea at the outset; instead the model emerged and evolved over time and via interaction. The initial vague model was regularly reviewed and discussed amongst a growing number of stakeholders and took shape as a result of these interactions. It follows the typical pattern suggested by the 'lean start-up' model of probing and learning through frequent interaction with the environment – albeit at the early stages this was a trial and error process rather than having a structure (Ries, 2011; Blank, 2013). But as the venture developed, so the use of a formal framework helped facilitate this since it made explicit many assumptions. This is important in the context of responsible innovation (RI) since it suggests that tools of this kind can help develop both reflexiveness and flexibility in the early design.

Analysis of the case also identifies a number of key points at which the innovation concept 'pivoted'. In particular, these pivot points gave rise to a number of interventions in the design of the innovation. We can think of a malleable innovation design space in which things are not yet established but where different stakeholder perspectives can and do shape the innovation. This suggests that early stage innovation has a strong potential for RI considerations if relevant tools can bring the key questions into the discussion.

It was also possible to identify a trade-off underpinning the pivots. In the early stages the challenge is around shaping the idea into something which actually works, using early market testing as a way of 'hunting' towards an optimum. But later pivots were more concerned with underlying values and beliefs; the original entrepreneur had deep views about the nature of the project but others were concerned about his business's viability. This is perhaps a classic version of the 'heart versus head' challenge facing social entrepreneurs – trying to do good may force uncomfortable compromises in order to maintain viability and long-term sustainability (Bessant and Tidd, 2015). In this case, two complementary models emerged: one well-fitted to its funding environment and run by medical staff largely for medical staff, and the other, operating on a more precarious funding regime, concerned with a bigger social agenda around empowering patients.

Responsible innovation raises questions familiar in the world of inclusive design and user participation – how to improve design quality and acceptance by early and active engagement of a wide range of stakeholders. The principles behind doing so are well-understood but there is still a need to explore and develop suitable tools and techniques to enable this to take place. The risk is that these considerations will be brushed aside in the interests of speed and focus but, as the case described shows, there may be considerable value in more systematic early stage exploration to avoid downstream problems. Structured frameworks which articulate key questions and explore different perspectives in systematic fashion can provide a useful addition to the entrepreneur's toolkit.

ACKNOWLEDGEMENT

This chapter is based on the results of the project 'Digitalize or Die – Dynamic Drivers of Responsible Research and Innovation in Health and Welfare Services' and is funded by Norwegian Research Council, project number 247716/O70.

REFERENCES

Abernathy, W. & Utterback, J. (1978). Patterns of industrial innovation. *Technology Review* 80(7), 40–47.

Baskerville, R. & Myers, M. D. (2009). Fashion waves in information systems research and practice. *MIS Quarterly*, 33(4): 647–662.

Bate, P. and Robert, G. (2006). Experience-based design: from redesigning the system around the patient to co-designing services with the patient. *Quality & Safety in Health Care*, 15, 307–310.

Bessant, J. & Buckingham, J. (1993). Organisational learning for effective use of CAPM. *British Journal of Management*, 4(4): 219–234.

Bessant, J. & Maher, L. (2009). Developing radical service innovations in healthcare: the role of design methods. *International Journal of Innovation Management*, 13(4), 555–568.

Bessant, J. & Tidd, J. (2015). *Innovation and Entrepreneurship*. Chichester: John Wiley and Sons.

Bessant, J., Abu El-Ella, N. & Pinkwart, A. (2015). Changing change management. In H. Albach, H. Meffert, A. Pinkwart and R. Reichwald (eds), *Management of Permanent Change* (pp. 105–120). Heidelberg: Springer.

Bessant, J., Smith, S., Tranfield, D. & Levy, P. (1992). Organisation design implications of computer integrated technology. *International Journal of*

Computer Integrated Manufacturing. Special Issue on Human Factors (July), 169–182.

Bevan, H., Robert, G., Bate, P., Maher, L. & Wells, J. (2007). Using a design approach to assist large-scale organizational change: '10 high impact changes' to improve the National Health Service in England. *The Journal of Applied Behavioral Science*, 43(1), 135–152.

Blank, S. (2013). Why the lean start-up changes everything. *Harvard Business Review*, 91(5), 63–72.

Bødker, K., Kensing, F. & Simonsen, J. (2004). *Participatory IT design: Designing for Business and Workplace Realities*. Cambridge MA, MIT Press.

Brabham, D., Ribisi, K., Kirchner, T. & Bernhardt, J. (2014). Crowdsourcing applications for public health. *American Journal of Preventive Medicine*, 46(2), 179–187.

Braverman, H. (1974). *Labor and Monopoly Capital: The Degradation of Work in the Twentieth Century*. New York: NYU Press.

Brown, T. (2008). Design thinking. *Harvard Business Review*, 84–92.

Buchanan, D. A. & Bessant, J. (1985). Failure, uncertainty and control: the role of operators in a computer integrated production system. *Journal of Management Studies*, 22(3), 292–308.

Crisp, N. (2010). *Turning the World Upside Down – The Search For Global Health in the 21st Century*. London: Hodder Education.

Eason, K. (1988). *Information Technology and Organisational Change*. London: Taylor & Francis.

Evans, P. & Wurster, T. (2000). *Blown to Bits: How the New Economics of Information Transforms Strategy*. Cambridge, MA: Harvard Business School Press.

Habicht, H., Oliveira, P. and Shcherbatiuk, V. (2013). User innovators: when patients set out to help themselves and end up helping many. *Die Unternehmung*, 66(3), 277–294.

Harhoff, D. & Lakhani, K. (eds) (2016). *Revolutionizing Innovation: Users, Communities, and Open Innovation*. Boston, MA: MIT Press.

Herstatt, C. & von Hippel, E. (1992). Developing new product concepts via the lead user method. *Journal of Product Innovation Management*, 9(3), 213–221.

Leonard-Barton, D. (1988). Implementation as mutual adaptation of technology and organization. *Research Policy*, 17, 251-267.

Moore, G. (1999). *Crossing the Chasm; Marketing and Selling High-tech Products to Mainstream Customers*. New York: Harper Business.

Morris, L., Ma, M. & Wu, P. (2014). *Agile Innovation: The Revolutionary Approach to Accelerate Success, Inspire Engagement, and Ignite Creativity*. New York: Wiley.

Mumford, E. (1979). *Designing Human Systems*. Manchester: Manchester Business School Press.

Osterwalder, A. and Pigneur, Y. (2010). *Business Model Generation: A Handbook for Visionaries, Game Changers, and Challengers*. New York: John Wiley.

Owen, R., Bessant, J. & Heintz, M. (eds) (2013). *Responsible Innovation: Managing the Responsible Emergence Of Science And Innovation In Society*. Chichester: John Wiley & Sons.

Pickles, J., Hide, E. & Maher, L. (2008). Experience based design: a practical method of working with patients to redesign services. *Clinical Governance: An International Journal*, 13(1), 51–58.

Ries, E. (2011). *The Lean Startup: How Today's Entrepreneurs Use Continuous Innovation to Create Radically Successful Businesses*. New York: Crown.

Rogers, E. (2003). *Diffusion of Innovations*. New York: Free Press.

Stilgoe, J., Owen, R. & Macnaghten, P. (2013). Developing a framework for responsible innovation. *Research Policy*, 42(9), 1568–1580.

Tidd, J. & Bessant, J. (2018, forthcoming). *Managing Innovation: Integrating Technological, Market and Organizational Change*. Hoboken, NJ: John Wiley.

Trist, E. & Bamforth, K. (1951). Some social and psychological consequences of the longwall method of coal-getting. *Human Relations*, 4(3), 3–38.

Utterback, J. (1994). *Mastering the Dynamics of Innovation*. Boston, MA: Harvard Business School Press.

von Hippel, E. (1988). *The Sources of Innovation*. Cambridge, MA: MIT Press.

von Hippel, E. (2005). *The Democratization of Innovation*. Cambridge, MA: MIT Press.

von Hippel, E. (2016). *Free Innovation*. Cambridge, MA: MIT Press.

Womack, J. and D. Jones (1996). *Lean Thinking*. New York: Simon & Schuster.

8. Responsible research and innovation: innovation initiatives for positive social impact

Raj Kumar Thapa and Tatiana Iakovleva

8.1 INTRODUCTION

> We believe that if we can create value to the society at large, and do our job well,
> satisfactory economic results will follow and allow us to build a stronger company
> with time.
>
> Åsmund S. Laerdal

Business organizations invest in innovation for the purpose of competitive advantage from corporate profitability and organizational growth perspectives. The primary focus of these organizations is in shareholders' value maximization, which requires business growth and quick turnover. Due to the growth imperatives and growing competition, the organizations are in constant pressure to invest in research and innovation for faster outputs (Christensen and Raynor, 2003).

Over the past five decades, business organizations have made substantial contributions in economic development and human wellbeing with innovation (Fagerberg, Fosaas and Sapprasert, 2012; Martin, 2012). However, with innovation and business activities, a number of societal, ethical and environmental issues have also been summoned in the course of development (Martin, 2016).

Despite the adoption of corporative social responsibility as their strategic management since quite a long-time, larger corporations are being criticized for not contributing enough to social and environmental reconstruction (Lazonick, 2014). Furthermore, due to the dominant shareholder value maximization ideology, these corporations are mainly concentrated in value extraction rather than value creation as per the interest of shareholders (Lazonick, Mazzucato and Tulum, 2013; Lazonick 2014). Such attitudes and practices have actually widened the societal and environmental problems.

These incentives of business raises the responsibility issues of the businesses, and raises the public concerns about the purpose and underlying motivations behind their business innovation (Owen et al., 2012).

Innovations which fail to address public concern and reflect on the underlying purpose, could meet public resistance (Asveld, Ganzevles and Osseweijer, 2015). Such innovations, though, possess the potential of addressing societal problems, and could potentially be responsible products, but would not be accepted by the users. Aligning with this line of argument, it is therefore necessary for the organizations to reflect on the purpose, process and outcomes of the innovation projects (Owen, Bessant and Heintz, 2013; Stahl et al., 2017). Such innovation outcomes could find diffusion to scale alleviating societal and environmental problems, thus increasing positive social impact.

With an explorative case study of a business organization within the medical industry, this chapter aims at answering the research question 'How do business organizations pursue responsible innovation in business development and create positive social impact?'

In the sections that follow, we present Responsible Research and Innovation (RRI), and social impact in section 8.2, followed by the research approach and details of the case company in section 8.3. In section 8.4, we present our findings followed by discussion and conclusion in section 8.5.

8.2   RRI AND SOCIAL IMPACT

Innovation is considered as a mechanism for competitive advantage to expand economic horizons (e.g., Teece, Pisano and Shuen, 1997). Business organizations are therefore concentrated on innovation outputs rather than innovation outcomes and the social impact of such innovations in society (Martin, 2016). Innovation has transformed the society and wellbeing. However, in many occasions, it is being exploited by certain interest groups at the cost of the rest of society and the environment. The growing economic disparity, environmental and ecological degradation are some of the alarming issues of recent times. Such issues raise public concern about the fundamental purpose, product, and outcomes and overall underlying motivations of such innovation (Stilgoe, Owen and Macnaghten, 2013; Stahl et al., 2017). This necessitates the focus of innovativeness of businesses which should be on innovation outcomes for broader social impacts rather than outputs as a means of competitiveness for economic growth.

'Social impacts include all social and cultural consequences to human populations of any public or private actions that alter the ways in which people live, work, play, relate to one another, organize to meet their needs, and generally cope as members of society' (Burdge and Vanclay, 1995, p. 59). Social impact of innovation therefore could be described as any change that particular innovation brings in the environment and society.

The principle aspect of RRI is responsible process for responsible innovation outcomes (Owen, Bessant and Heintz, 2013). This implies that business innovation depends not only on the novelty of the product or services that an organization offers to the customers, but also on how they produce and deliver, and what the potential impact of such a product or service could be on society at large.

This also implies that innovations targeted at alleviating societal and environmental problems are diffused to a scale thus benefiting larger communities by creating a positive social impact. The focus of businesses should therefore be on *purpose*, *process* and *outcomes* of their research and innovation activities, as suggested by Stahl et al., 2017 in their maturity RRI model.

8.2.1 Purpose

The purpose of a research and innovation project must be clearly specified. Organizations can reflect on the purpose of their research and innovation project, for instance through a mission statement. The purposed statement reflects the interest and motivation for a particular research and innovation project. In order to cast a positive social impact, the value creation through research and innovation should be aligned with the norms, values and legal status of society. The purpose of such research and innovation should be emphasized, based on the need of society, for instance alleviating societal problems keeping in mind to align the innovation activities as per the expectations of society and not just avoiding harm. The purpose should be focused mainly on societal desirable innovation process targeting at responsible outcomes.

8.2.2 Process

The process is about reflection on action. What are the necessary actions organizations should perform in order to align with their purpose of research and innovation? Who are included in the research and innovation activities and for what purpose? For example, inclusion of users and the public will bring diverse knowledge in the innovation process. In many occasions users have solutions to the problem, while on other occasions they have the solution but may not have the know-how to realize the product. Further, on some occasions they need to be educated, made familiar and even empowered. It is important to define clear outcomes while engaging stakeholders into innovation process. This represents responsible thinking and responsible acting by the organizations, which helps them to add social impact.

8.2.3   Outcomes

The outcomes of research and innovation cannot be confirmed at the beginning of the process. In some occasions, though the purpose and motivation of innovation is for responsible outcomes and the responsible processes are followed strictly, the outcomes may result in causing negative externalities in society and the environment. In such a situation, the essential act is to respond to such consequences immediately and avoid causing further damage. It is to be acknowledged that RRI processes will not necessarily bring responsible outcomes, but help develop a culture of reflecting and responding to take care and protect the society and environment from further damage. The aspiration of RRI can be achieved only if an organization demonstrate responsiveness, meaning either abandoning manufacturing a product which adds negative externalities in society and the environment or take it back to discussion for modification or an alternative solution. Such acts enhance reflexivity of the organization and demonstrate how serious they are about taking care of the environment and society. This could enhance trustworthiness of the business organizations in society, which could affect reputation and brand image.

Innovation needs to be diffused to scale in order to benefit larger communities and scale up the positive social impact in society. Innovation cannot be considered responsible until and unless it brings positive change in the society (von Schomberg, 2013).

Scaling up can be done through depth and breadth. Scaling depth here means to spread responsible innovation deep into the community at local and national levels, while scaling breadth means spreading responsible innovation out of the national boundary so that the global community can access such an innovation in alleviating their local or regional problems.

8.3   RESEARCH APPROACH

In the current study, we adopted a single case study method (Yin, 2003) of a Norwegian company operating in the health industry. Data collection involved in-depth interviews performed by the authors with a technical director associated to Laerdal Emergency Care, a project manager of Laerdal Medical and the director of educational strategy and standards in partner company 'SAFER' (Stavanger Acute Medicine Foundation for Education and Research). In addition, we used observations (visits to the production facilities of Laerdal Medical) and secondary material available in the public domain, including the Laerdal Report on Sustainability (2016), the Laerdal Impact Report (2018), the Laerdal website (Laerdal, 2018) and a book about Laerdal, *Saving More Lives Together – Vision for 2020* by Nina Tjomsland (2015). Besides these, two student reports devoted to innovation processes in Laerdal, based on three interviews performed in 2018 (recorded and transcribed) with the supervision of authors are also included as additional materials.

8.3.1   The Case Company

Laerdal AS was founded back in 1940 by Aasmund S. Laerdal, in Stavanger, a small city in Norway. Inspired to bring joy for local kids, his business venture was focused on commissioning and publishing books and manufacturing wooden toys for kids. From the beginning of the establishment, the focus of the company was in constant product development. For this purpose, Aasmund shared his visions with the employees, and he searched out the professionals and experts who could assist and advise him. He invested in education, explored different nations and cultures, and brought back knowledge, which could assist him to achieve his vision and sought out new opportunities. Back in 1949, he came to know the use of soft plastic in manufacturing toys for kids. With constant passion, determination, dedication and experimentation, he finally achieved success in manufacturing his first toy, Anne, from soft plastic in 1950 and his company became the first to manufacture dolls out of soft plastics in the European region. This expertise was later extended in manufacturing the popular Tomte toy cars. The company saw an opportunity for mass production and export to low-cost countries and was able to export toys to over 110 countries.

From kids' toys to life saving
Since the establishment of the enterprise, Aasmund was constantly thinking about a new product development, a product that was suited for production in Stavanger. Being an expert in soft plastic in Norway, Laerdal was approached by Civil Defence in 1953 and asked to develop imitation wounds for training. This brought an opportunity for Laerdal to extend its network into the medical profession. With its expertise in manufacturing dolls and cars out of soft plastics, and collaborating with the physicians and experienced surgeons, it was able to create accurate models of relevant wounds. Furthermore, Laerdal was able to produce a glue to fasten the wounds on the skin without hurting while subsequently removing it after constantly testing different types of glue.

In 1954, a personal incident happened with the firm founder. Aasmund found his two-year-old son floating unconscious in the sea. He immediately took the inert body of his son out of water, slapped and shook him to clear his airways, and stimulated him to breathe, thus saving him. This emotional incident laid the foundation for the company's next major focus area.

With the extended networks with health specialists, Aasmund came to know that there had been substantial development in a new and much better method of resuscitation, involving mouth-to-mouth breathing by a group of physicians and engineers in Baltimore.

The emotional incident that Aasmund experienced while saving his own son and the new development within resuscitation inspired him to enter into the manikin manufacturing business. Making manikins was difficult, but the more difficult part was how to overcome psychological and cultural resistance among people to make them acceptable to use. Beside these issues, price and quality of the product should not be ignored in order to make many people benefit from such innovation. In close collaboration with anaesthetists in Stavanger, Aasmund was successful in prototyping his first manikin, 'Resusci Anne' in 1960 and he demonstrated it to American resuscitation pioneers in New York. This event resulted in friendship and the collaboration of Aasmund with one of the resuscitation pioneers, Peter Safer, who suggested the inclusion of a chest ring for compression training, which Laerdal Medical company complied with. The still looming challenge at that moment was about inspiring as many people to learn and implement the mouth-to-mouth technique to help save lives.

Being a strong advocate of resuscitation, an anaesthesiologist Bjorn Lind, played an essential role in bringing his colleagues to join forces to convince people of the importance of mass training in resuscitation. The first and major breakthrough occurred when a group of banks donated 650 manikins to primary schools where both teachers and students were empowered in using manikins. Dr Lind followed the training and noticed that children learned resuscitation well, like their teachers (Lind, 1961). Furthermore, the outcomes of such intervention were scientifically recorded, analysed and made available (Lind and Stovner, 1963). Laerdal Medical was then able to attract international attention, and Norway emerged as a pioneering nation in the history of life-supporting first aid and became a role model in conveying the message to the world that every schoolchild can learn how to save a life.

Aasmund's obsession of helping save more lives intensified in the search for more opportunities. The company was proactively involved in healthcare innovation. In August 1961, in close cooperation with Dr Safer and German specialists, Laerdal Medical initiated and hosted the First International Symposium on Emergency Resuscitation in Stavanger, which was attended by specialists from all over the world.

The establishment of a Cardio-Pulmonary Resuscitation (CPR) committee by the World Federation of Anaesthesiologists in 1964 and a discovery that conveyed that external chest compressions could provide a circulation of blood to the brain when the heart stopped beating, and increase greatly the possibility of revival, proved to be crucial potential for the Laerdal company. As a result, in 1969, Laerdal Medical produced Resusci Anne for CPR, capable of being used to practice artificial ventilation and external chest compressions. In addition to this, a series of other products associated with first aid were produced and Resusci Baby was one of them. The same year was marked with the introduction of the Resusci Folding Bag, Pocket Mask to protect the rescuer and Vacuum Mattress to protect the patient, followed by a disaster kit.

In 1971, Laerdal Medical introduced Resusci Anne, equipped with a printer giving feedback to the trainee and at the same time providing important information about the efficiency of the training and possible areas of improvement in the manikin.

Following the recommendation of the American Heart Association (AHA), Laerdal Medical introduced teaching of CPR to lay persons in 1973 and that proved to be the big step in empowering lay people to help save more lives. To promote teaching of CPR, Laerdal Medical printed informational material for the medical sector in 15 different languages.

In 1978, Laerdal Medical decided to concentrate fully on saving lives and stopped producing toys for children. In the same year, Aasmund received an International Award of the AHA and became the first non-physician to receive such award. The same year he became an honorary member of both the British Association for Immediate Care and the Norwegian Society of Anaesthesiologists. He was also honoured by the University of Pittsburgh.

By 1979, Laerdal Medical was the established market leader exporting abroad 95 per cent of the outputs from several production lines. The Laerdal company decided to channel some of the profits into a new foundation, hence in 1980 the Laerdal Foundation for Acute Medicine was established in collaboration with the University of Oslo. The foundation carries out research projects and educational initiatives.

The responsibility of running the company transferred to Tore Laerdal, after the death of his father Aasmund in 1981. Since then, innovation for broader impact has become Laerdal's culture. In addition to this, Laerdal is actively involved in any innovation activities in association with helping to save lives. In 1982, the Laerdal Foundation helped initiate and support an international conference for CPR trainers in London together with the AHA. During the 1980s, Laerdal collaborated on two projects for mass CPR training in the Stavanger region under the motto 'Action Rogaland 1983: You can save lives'. This initiative resulted in 5,000 volunteer learners over just two weekends.

Inspired by such impressive outcomes, Stig Holmberg from Gothenburg developed a Swedish CPR training model to train all healthcare personnel. Laerdal contributed to the training programme and also sponsored thousands of posters to hospitals and health institutions. The posters were designed by Laerdal and were simple to understand and easy to learn from.

Since the early 1990s, posters for both basic and advanced life support were printed in many languages for European Resuscitation Council and were displayed in thousands of hospital emergency rooms and training sites.

Beside these supportive activities, Laerdal constantly focused on innovation for product efficiency advancement for better outcomes and impacts. In 1999, with close collaboration with physicians, Laerdal developed SimMan, an advanced patient simulator. The development of SimMan was to help the extensive training of health personnel in US hospitals in order to reduce an estimated 50,000–100,000 of unnecessary deaths each year due to errors made in those hospitals.

In the same year, Laerdal bought Medical Plastics Laboratory Inc. (MPL) in Gatesville, Texas and introduced the SimMan project (Tjomsland, 2015). MPL is now called Laerdal Medical Texas where SimMan is the major production. Since its entry into the medical industry, Laerdal Medical developed a broad range of products and programmes to support resuscitation training and emergency interventions. With a focus on increased patient safety, in 2000, Laerdal Medical broke ground in the field of medical simulation with the introduction of relatively low-cost patient simulators, allowing for risk-free interactive training in emergency care.

Laerdal was aware that helping to save lives is not possible all alone despite a continuous focus of development and innovation. It continuously searched for collaborators and partners with whom values, expectations and purpose could be aligned with its core mission of helping to save more lives. In addition, Laerdal constantly looked for opportunities to acquire the companies associated with its core mission. In 2002, Laerdal collaborated with a Danish-based company, Sophus Medical to further explore interactive medical training products. In 2003, Laerdal acquired Sophus Medical, which is now called Laerdal Medical Sophus and is leading in the field of micro simulation training.

During 2004, Laerdal was able to launch extensive products covering educational micro simulation programmes for pre-hospital, in-hospital and military segments. In the same year, Laerdal officially opened a new factory in China which enabled it to provide a quality facility in the East Asian region. Similarly, it opened a factory in Monterrey, Mexico, in 2006.

In 2006, Laerdal's attention was drawn towards maternal, newborns and child health. It is highly acknowledged by the American Academy of Paediatrics (AAP) that the heart of simulation lies neither with the simulator nor the technology, rather with the educational methodology. Laerdal partnered with AAP to advance educational science and resources necessary for training in neonatal resuscitation.

Simulation-based training in neonatal resuscitation programmes was being used in 120 countries and was in high demand. However, it was complex and resource-demanding to implement such programmes in low-resource countries. Unfortunately, the need for such programmes is high in those low-resource countries where many children and mothers lose their lives during birth due to lack of birth-related education and training. Laerdal was looking for ways to address such challenges in low-resource countries. As a result, in 2010, a daughter company, Laerdal Global Health (LGH) was established as a not-for-profit company in order to develop high-impact, low-cost training and therapy products aimed at helping save the lives of newborns and mothers in low-resource countries.

Today, Laerdal is a global company with more than 1,500 employees in 24 countries, dedicated to helping save lives through resuscitation, emergency care, and patient safety.

With its own expertise and existing networks and collaborations, the company targets at scaling its social impact globally by helping save more lives through constant improvement and innovation.

8.4 FINDINGS

8.4.1 Purpose

The Laerdal company is driven by its mission of helping save lives. 'Help save half million additional lives every year by 2020' is the new goal that the Laerdal company announced on their web-page and in strategy documents. This ambitious goal underscores the need for focusing the activities and organizational capabilities on areas where Laerdal believes it can make the biggest impact. With a long experience and expertise within resuscitation research, patient safety, and global health initiatives, Laerdal is confident in achieving its goal. The company's mission statement of helping to save more lives reflects the purpose of organization, research, and innovation that the company carries out and supports. Though profit is essential for scaling up innovation capabilities in order to scale up social impact, it is not the top priority of the company.

8.4.2 Process

In order to achieve the ambitious goal and scale up social impact, Laerdal is continuously engaged in innovation aligned with its core mission, helping save more lives. For Laerdal, innovation is not about the bare novelty of technologies or products but it is about its contribution to helping save lives. This implies innovation in Laerdal is about impact, and innovation can only have enormous impact if it is diffused to scale. This requires the detailed understanding of needs and expectations of end users.

As expressed by the founder, Aasmund Laerdal, "sustainable business is only possible through developing [the] ability to listen, endless curiosity, practical problem solving, respect for [the] customer, hard work and a passion for continuous improvement" (Tjomsland, 2015, p. 16). This has translated into an organizational culture within Laerdal. Meeting customers' and users' needs and expectations requires their inclusion at the very early stage of the innovation process. Laerdal articulates its sales forces located around the globe to include the users and customers' voices and ensures their voices are heard in the innovation process.

> We gather constant customers and users feedback about our product and services through our sales forces located all over the world and constantly try to respond to the feedback and their expectations whenever and wherever possible. *Technical director*

Achieving ambitious goals and attaining sustainable business growth is not possible without reliable collaboration, partnerships and networks. Laerdal has succeeded in collaborating with a number of complementary organizations that can build and implement solutions for end users. At the same time, it is constantly seeking to extend networks with reliable partners with common goals and shared values. The company believes that future success can only be realized through strong commitment to global partnerships, cooperation, and constantly striving to further develop alliances and partnership in many countries. Establishing strong and sustainable cooperation requires constant demonstration of reflexivity and responsiveness on actions and commitments. Through its responsible attitude and actions, Laerdal is able to bring many strategic partners on board within its network to support its mission of helping save more lives.

As of today the Company collaborates with The American Heart Association (AHA), the American Academy of Pediatrics (AAP), HealthStream, the National League for Nursing (NLN), Philips Healthcare, SAFER (Stavanger Acute medicine Foundation for Education and Research) and Jhpiego. With close collaboration and alliances with these strategic partners, Laerdal is able to innovate quality products and services in order to enhance helping save more lives. These collaborations actually formed a platform for constantly learning for better innovation outcomes and even co-creating innovative solutions, for instance, Laerdal collaborates with AHA on several large-scale projects such as CPR, a CPR in schools programme, Heartcode eSimulation courses and Resuscitation Quality Improvement Programme (RQI) being able to revolutionize resuscitation. With close collaboration with APP, Laerdal co-created several simulators supporting the Newborn Resuscitation Programme (SimBaby, SimNewB, and Premature Anne), e-learning programmes, and the suite of Helping Babies Survive educational modules. Laerdal's collaborators are also customers and users of the majority of the Laerdal products. Working closely with them therefore enabled Laerdal to understand actual needs and expectations. To meet their expectations and establish strong relationships, the company focused on innovation management from the design stage onwards.

> We prototype many different products and test them with customers and users. If one fails, the other would work. If all fail, we redo it with feedbacks from customers, users and the experts until we get satisfactory outcomes. *Project manager*

> Keeping customers' and users' satisfaction high to build sustainable relationships is what Laerdal believes. Responding to the constructive feedback of the users and customers is to demonstrate respect and proof that their voices are heard and respected.

> Some of our customers did not like colour of the manikins and asked us to change it to represent different races. Though it is costly and difficult procedures for us, we do respect their cultural expectations and changed as per their suggestions. *Project manager*

Networking with external partners is also forming a platform for inclusion of diversity of knowledge and expertise, which in fact enables Laerdal to understand the desirability and acceptability of innovation in different contexts. Including Jhpiego, a non-profit health organization affiliated with the Johns Hopkins University, Laerdal is able to successfully implement the Helping Mothers Survive Programmes in low-resource countries.

In order to focus on need-based research and innovation to achieve its ambition, Laerdal company keeps itself updated with the most recent research associated with healthcare. Besides its own research centre, Laerdal contributes to research through donations to the Laerdal Foundation for Acute Medicine and support to the SAFER simulation centre.

8.4.3   Outcomes

Innovation outcomes are uncertain. However, adequate consultation with users, customers, stakeholders, experts and other relevant actors at the beginning of the innovation process could lead to better outcomes. In addition to responding to the feedback from the stakeholders, customers, users, and partners, they need to keep room for constant improvement of the product and services for better productivity. In other words, the probability of achieving expected innovation outcomes is higher when innovation activities are aligned with purpose.

The revolution on resuscitation, emergency care, and helping babies survive and helping mothers survive is only possible through responsible innovation outcomes that Laerdal has adopted and implemented. For instance, the e-learning programme and initiatives such as Resuscitation Quality Improvement (RQI) are innovation outcome that have optimized the way

CPR training is delivered. Simulation training and related activities can be considered as another responsible innovation outcome, which contributes in the reduction of deaths due to medical errors.

8.4.4 Innovation Diffusion – Scaling Depth/Breadth

As the historical development of the company illustrates, Laerdal Medical was quite successful in offering its product and services within treatment of sudden cardiac arrest and emergency care in high-resource countries. The company has a broad international market, offering its products and services in 24 countries. Since 2010, via LGH, the company has extended its services in low-resource countries in Asia and Africa in order to help save lives of newborns and mothers, preventing deaths due to birth-related complications. In one particular hospital, over the period of six months, the application of Laerdal training through the instalment of modern baby manikins combined with a digital self-assessment training programme, resulted in reducing the baby death rate by 40 per cent. It can be concluded that the company is constantly scaling up innovations both in terms of depth as well as breadth.

At the same time the company is conscious with regard to how and where to grow its business. Aligning its growth ambitions with its mission of saving lives lies in the core of the growth strategy for Laerdal. The company is trying to achieve growth through building networks, alliances and partnerships with trusted partners to achieve higher social impact.

> We cannot grow as we want because we need to concentrate on our capacity, quality and maintain our image at the same level as we have been able to so far. Focus on growing faster would be problematic for maintenance of organizational culture and overall performances. The only thing we could do is to find the right partner and build alliances. *Technical director*

We summarize the major RRI elements, such as purpose, process, outcomes as well as scaling and social impact for Laerdal company in Table 8.1.

The company can scale up positive social impact, further increasing its RRI approach by covering resuscitation and emergency care services in low-resource countries. For this, there should be local and national government initiatives and healthcare policies need to be considered, and this clearly signifies the need for inclusion of relevant stakeholders and actors from respective societies.

Table 8.1   *RRI approach and social impact*

Purpose	Process	Outcomes	Scaling	Social Impact
Stated clearly in mission statement: Help save more lives (saving lives through quality and accessible healthcare facilities)	Research and innovation aligned with the core aspects of RRI (inclusive decision in research and innovation activities) Production of training and educating materials (need-based innovation)	Quality products (increased efficiencies as per the expectations of customers and users) Empowering users and local community for quality healthcare Educating and empowering school kids in quality CPR, providing kits necessary for it	Production sites extended in global scales Introduction of new business unit and innovative approach to subsidize costs for low-resource countries Extending programmes to support helping save more lives mission	Saved more lives (reduced death due to sudden cardiac arrest and deaths due to medical errors in high-resourced nations) Saved lives of newborns and mothers in low-resource nations (reduced death of newborns and mothers due to birth-related complications) Extended affordable care services in low-resource nations in Africa and Asia

8.5   DISCUSSION AND CONCLUSION

Drawing on the case study of the Laerdal company, this chapter has presented how socially responsible attitudes and activities of organizations during business innovation can actually boost positive social impact and sustainable business growth.

As a known proverb says: 'Necessity is the mother of invention'. Demand-based innovation or user innovation developed with broader inclusion of stakeholders and lay people are more compatible and bear higher probabilities of innovation diffusion to scale. Furthermore, such bottom-up innovation initiatives not only facilitate the development of innovations, but also contribute to social empowerment (von Hippel, 2005, 2017). However, inclusion of stakeholders, users, and the public in the early stage of innovation activities as per the aspirations of RRI appears to be a costly, time consuming and complicated process. Such approaches can easily be presumed to result in delayed decision-making or the state of no decision at all. Such perspectives prioritize monetary interests over and above the interest of the stakeholders, customers and society, which would not be beneficial for businesses in the long run. In fact, underestimation of the role of such societal actors would be too costly for the companies. Inclusion of more people, especially user groups at the very beginning of the innovation process, gives deep insights into better solution designs and responsible innovation outcomes. Such inclusive and responsible aspects have enabled Laerdal to be the world leader in resuscitation and medical simulation. A socially responsible purpose and organizational commitment are the major factors behind Laerdal's success.

Stating an ambitious, purposeful and impressive mission statement does not itself reflect responsible behaviour of a company. Such an impression will be illusive if the company fails to translate the statement into actions. This is also associated with companies' trustworthiness and reputation in society, which will ultimately determine the survival and growth of the business. Socially responsible business, in the long run, can be profitable by being responsible. This line of our argument is aligned with the case of Laerdal, which is able to extend its mission of helping save more lives to the next level, aiming at helping save 500,000 more lives every year by 2020 (Laerdal Report on Sustainability, 2016), extending its positive social impact in society. We therefore argue that business innovation aligned with RRI aspirations would enable us to address social problems in an efficient way. Such initiatives from business communities mean sustainable business and positive social impacts of their business innovation to society.

Innovation that incorporates public concerns about the purpose and motivation of innovation would get a green signal from society for desirability and acceptability to a larger extent, since such innovations are designed and developed for society, with society (Owen, Macnaghten and Stilgoe, 2012). Furthermore, such innovations carried out with broader inclusiveness, representing public concerns, values and expectations would result in diffusion to scale, which is necessary for creating and extending social impact. Being able to cast positive social impact in society, facilitates building trust among the stakeholders and the public. This is how organizations can build brand image and get continuous support from society. This in fact adds a moral obligation of organizations to maintain their image in the public domain and opens up growth opportunities to spread positive social impact further. This, in turn, extends the level of social responsibility of businesses, an obligation of taking care of society and the environment, a culture of responsible business innovation.

Our line of argument here is responsible innovation, activities aligned with purpose, process and outcomes would result in socially acceptable or even desirable innovation outcomes. Such innovations are compatible and achieves diffusion to scale creating social, environmental and economic values. However, how to ascertain that responsible acts leads to responsible outcomes remains ambiguous. Social impact assessments with respect to RRI could be one possibility to evaluate how far companies are able to extend positive social impact through business innovation activities.

Based on analysis of our case firm, its research and innovation process, purpose and outcomes and literature we therefore suggest the following Social Impact Assessment framework (SIAF), depicted in Figure 8.1.

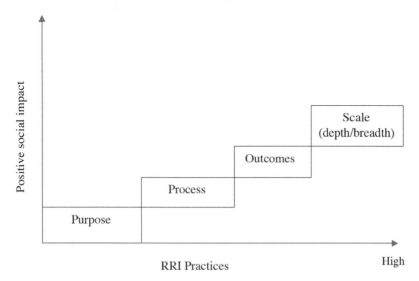

Figure 8.1   *RRI and social impact*

Social impact assessment frameworks (SIAF) can actually enable businesses to perform self-assessment of their own research and innovation projects and to evaluate the socioeconomic and environmental impact of such projects. The SIAF will not only find application in assessing responsible innovation activities within the organization but also facilitate organizations to build their reputation and sustainable growth opportunities. What we have demonstrated through this study is that prioritizing societal interest over mere profitability would actually result in both profitability and sustainable growth in the long run and this is associated with the socially responsible purpose and outcomes of business innovation. However, profitability of business should not be overlooked. Profitability is essential for research and innovation of products or services and supplying for broader communities for exponential positive social impact. We therefore emphasize the necessity for businesses to align with purpose, process and innovation outcomes.

RRI thinking and practices in research and innovation would enable businesses to identify and prioritize the most relevant innovation necessary for societal advancement. Such practices help in initiating innovation culture to focus need-based innovation and innovation aligned with the interest, value and legal status of society and keep aware of not adding negative externalities which could mean the cause of social and environment destruction. RRI practices would enable all societal actors to change their mindset and enable them to take care of society and the environment, or at least demotivate them from performing any actions, which could destabilize society and the environment. Any initiative to take care of the environment and society would definitely leave a good message in society and cast a positive social impact.

However, a responsible process, product and responsible outcomes by themselves would not necessarily cast sustainable positive social impact unless responsible innovation is fully diffused in society for broader social reformation. This implies the necessity of scaling up such innovation to serve a global community. Alleviating local, regional or national problems does not necessarily represent eradication of the problem, since a problem in any part of the world be the cause of problem for the local community in the present context of globalization. Scaling up of positive social impact is only possible if an organization could scale up responsible innovation both locally as well as globally; scaling up in terms of depth and breadth. The underlying aspect of RRI is to orchestrate research and innovation activities necessary for maximizing positive innovation outcomes for positive social impact and not just on research and innovation activities to avoid harm.

REFERENCES

Asveld, L., Ganzevles, J. & Osseweijer, P. (2015). Trustworthiness and responsible research and innovation: the case of the bio-economy. *Journal of Agricultural and Environmental Ethics*, 28(3), 571–588.

Burdge, R. J. & Vanclay, F. (1995). Social impact assessment: a contribution to the state of the art series. *Environmental and Social Impact Assessment*, 14(1), 59–86.

Christensen, C. M. & Raynor, M. E. (2003). *The Innovator's Solution: Creating and Sustaining Successful Growth*. Boston, MA: Harvard Business School Press.

Fagerberg, J., Fosaas, M. & Sapprasert, K. (2012). Innovation: exploring the knowledge base. *Research Policy*, 41(7), 1132–1153.

Laerdal home page (2018). Avaiable at https://www.laerdal.com/gb/ (accessed 16 August 2018).

Laerdal Impact Report (2018). Impact update January 2018, https://cdn0
.laerdal.com/cdn-4a208f/globalassets/documents/our-company/laerdal
-impact-report.pdf (accessed 16 August 2018).

Laerdal Report on Sustainability (2016). Help build better society. http://
viewer.zmags.com/publication/954ac3b8#/954ac3b8/8 (accessed 16 August
2016).

Lazonick, W. (2014). Profits without prosperity: stock buybacks manipulate
the market and leave most Americans worse off. *Harvard Business Review*
(September), 1–11.

Lazonick, W., Mazzucato, M. & Tulum, Ö. (2013). Apple's changing business
model: what should the world's richest company do with all those profits?
Accounting Forum, 37, 249–267.

Lind, B. (1961). Teaching mouth to mouth resuscitation in primary schools.
Acta Anaesthesiol Supplement, 3, 63–66.

Lind, B. & Stovner, J. (1963). Mouth-to-mouth resuscitation in Norway.
JAMA, *185*(12), 933–935.

Martin, B. R. (2012). The evolution of science policy and innovation
studies. *Research Policy*, 41(7), 1219–1239.

Martin, B. R. (2016). Twenty challenges for innovation studies. *Science and
Public Policy*, 43(3), 432–450.

Owen, R., Bessant, J. & Heintz, M. (eds) (2013). *Responsible Innovation:
Managing the Responsible Emergence of Science and Innovation in Society*.
Chichester: John Wiley & Sons.

Owen, R., Macnaghten, P. & Stilgoe, J. (2012). Responsible research and innovation: from science in society to science for society, with society. *Science
and Public Policy*, 39(6), 751–760.

Stahl, B., Obach, M., Yaghmaei, E., Ikonen, V., Chatfield, K., Brem, A. ... &
Brem, A. (2017). The Responsible Research and Innovation (RRI) maturity
model: linking theory and practice. *Sustainability*, 9(6), 1036.

Stilgoe, J., Owen, R. & Macnaghten, P. (2013). Developing a framework for
responsible innovation. *Research Policy*, 42(9), 1568–1580.

Teece, D. J., Pisano, G. & Shuen, A. (1997). Dynamic capabilities and strategic management. *Strategic Management Journal*, 18, 509–533.

Tjomsland, N. (2015). Saving more lives-together. The vision for 2020. https://
laerdal.com/gb/about-us/ (accessed 27 November 2017).

von Hippel, E. (2005). *Democratizing Innovation*. Cambridge, MA: MIT
Press.

von Hippel, E. (2017). *Free Innovation*. Cambridge, MA: MIT Press.

von Schomberg, R.(2013). A vision of responsible research and innovation. In
*Responsible Innovation: Managing the Responsible Emergence of Science
and Innovation in Society*, R. Owen, J. Bessant, and M. Heintz (eds), 51–74.
Chichester: Wiley.

Yin, R. K. (2003). *Case Study Research: Design and Methods*, 3rd edn. Thousand Oaks, CA: Sage Publications.

9. The Blink innovation story – viewed through the lens of responsible innovation

Dagfinn Wåge and Andrea Marie Stangeland

9.1 INTRODUCTION

This chapter discusses a welfare technology concept called Blink. Blink is developed by The Research & Development & Innovation department within the Lyse Group. The Lyse Group is a Norwegian multi-utility, multi-service company with a legacy in renewable energy production. More than a century ago the company commenced hydroelectric power production and sales in Rogaland, south western Norway. In the early 2000s, Lyse diversified from being an energy provider to becoming an energy and communications provider. In 2001, the company commenced the rollout of fibre optic infrastructure that would provide consumers with TV, Internet and telephony services using Fibre-To-The-Home technology. The company continues to expand its communications offerings and has become a provider of smart home and smart energy solutions to consumers and prosumers. In 2018, more than 500,000 homes were connected using Lyse's fibre optic infrastructure. Within the space of two decades, the company has seen a rapid increase in revenue generation from its communications services. In 2018, Lyse's communications business unit generated the same level of revenues as its energy production business.

Lyse's research and development department continuously work on developing new technologies and finding new business areas for the company. Blink is one of the projects that Lyse was developing recently. Although Blink is a tried and tested concept, it is not available in the market. The Lyse Group is currently not active in the welfare technology sector.

Blink is a new and innovative video service concept. It combines Internet of Things (IoT) support with a Full HD video service supporting devices such as smartphones, tablets and TV sets. Although Blink has evolved to support a broad range of use cases, the initial premise that it should be used in an at-home care setting still holds true.

In this chapter the background for Blink, its development phases and the basic functionality of the product are discussed. The purpose of this chapter is to contribute to the discussion from the point of view of the industry practitioner. In retrospect, the development of Blink was examined through the lens of responsible innovation and placed in the context of the four basic components of responsible innovation: anticipation, reflexivity, inclusivity and responsiveness. Finally, a short summary of learnings from Blink is provided.

It should be noted that no one in the Blink-project organization have had any knowledge of these theories until now. You will also see that the use of literature references is absent in this chapter, as the whole purpose is to analyze a practical innovation project under the lens of responsible innovation.

9.2   BACKGROUND

This story started in 2009, where we gathered some of the top managers in Lyse to find what we referred to as megatrends. We started out with a technology focus, as technology evolves very fast. But we did not end up with technology-driven megatrends. On the contrary, we found something we called 'driving premises of the future' which we believe is much stronger than a megatrend, because they are bound to happen, and therefore do not contain the inherent uncertainty that we find in trends.

Three 'driving premises of the future' were identified. They were: (1) dramatic demographic change (the population will consist of a much higher number of elderly people), (2) environmental and climate change, (3) urbanization and its impact on rural areas.

In order to gain insight on the demographic challenge in the Stavanger region, two local municipalities were contacted. Discussions with municipalities on the topic of changing demographic profiles led to the development of a project entitled 'Cooperative development project within welfare technology' (FUIV-project). The starting point for this project was a number of anonymized accounts of everyday challenges faced by nurses who perform home visits.

Based on these accounts three user groups were identified: healthy elderly, elderly with early cognitive impairment and elderly with mobility impairment. Fifteen prototypes for new services were developed and installed in the homes of 20 elderly people who belonged to at least one of the three user groups. The services aimed to respond to individual user needs. Services were grouped into three main categories: safety services, smart home services and communication services.

Full HD video communication service via the user's home TV screen belonged to the latter category. The video communication service was sub-divided into three separate services: video communication with friends and relatives; video communication with nurses performing home visits and finally a Red Cross virtual visiting service using video. User testing was a fundamental part of the development process. A deaf man living in Bergen and his sister in Haugesund, 200 km south of Bergen, tested the video communication service. A mentally disabled man living in a nursing home in Oslo used the service to communicate with the persons living in his visiting home located in Stavanger. All these use cases gave us a lot of insight in how video could enrich lives and make work processes more efficient.

The video communication service quickly became popular amongst users. Unfortunately, the external video service (not the Blink service) terminated their product. User of the service were left without an alternative.

Norway is a sparsely populated country with many small communities spread along a long and often inaccessible coastline. Kvitsøy, an island with 530 residents, is the smallest municipality in Rogaland. Its residents are dependent upon ferry transport in order to reach services that are based on the mainland. Video communication solutions represents tangible benefits for citizens in isolated regions such as Kvitsøy. Video communication offers face-to-face communication without the need to travel. Chronic obstructive pulmonary disease (COPD) – patients on Kvitsøy participated in a trial that combined the use of sensors (oximeters and spirometers) with video on a PC contained in a plastic suitcase. The price of this product was very high and the solution was proprietary. Through communication with nurses performing home visits, it became clear that none of the COPD patients were willing to take the measurements unless a skilled nurse at the mainland hospital provided guidance via video. Therefore, the combination of IoT measurements and video was crucial. Combining this with the user's best screen – the TV – it became possible to lower the barriers of use and cost.

9.3   THE BLINK INNOVATION AND ITS TWO PHASES

Our first move was to search for a replacement product, but we were not able to find it. We even travelled to the Consumer Electronics Show (CES) in Las Vegas with no result.

The only way forward was to try to make the product ourselves. We were not able to do hardware design in Lyse, and we needed to obtain more competence in home care. The result was an open innovation project between three regional partners; Lyse, Westcontrol and Norsk Telemedisin.

9.3.1   Phase 1 – Prototyping and Beta Testing

This process started in 2014, and after having studied the use of video amongst different user groups, we decided that the basic design principle should be user-friendliness and focus on universal design. A product that is difficult to figure out or inefficient to use will not win much of a user base. When designing for the variety of users in the care segment, universal design is a good starting point. The design phase resulted in a lot of requirements, making the product more complex, but better for the end users. The main requirements were as follows:

- A user-friendly remote control that was tactile and had different LED-light colours that guided the users if in doubt. The goal was to give them a better understanding of the link between the possible interactions on the screen and how to perform the interactions using a remote control, since most TVs do not have a touch interface.
- An intelligent HDMI-switch, to avoid users to be confused regarding different input sources on the TV (this had been a challenge in the pilot). The intelligent HDMI-switch gave us the opportunity to provide video call as an overlay, and the only thing the users needed to do was to decide whether or not to answer the call.
- Avoidance of 'video-islands' where it was impossible to call between different vendors of video solutions. This was possible using software-based video-bridging provided by a third party.
- Adding different additional services like calendar functions, reminders, easy access to pictures and videos being sent by email directly on the TV.
- Finally, the product should be affordable.

We managed to put together a working prototype in January 2015, for which we received a CES Innovation Award at CES in 2015. But the prototype was a standalone product and did not have all the necessary infrastructure and back office functions in place.

In order to pilot test the Blink service a beta version of the product and platform was built. This development was time consuming and difficult choices regarding protocols, hardware components and architectural issues were made. The pilot tests included testing the interaction design on the user interfaces and that the main requirements were being met.

Following two years of development and testing, including user-testing in five homes, the conclusion was that the total solution was not stable enough for use. From time to time unexpected errors occurred such as poor lip sync and calls terminating during the setup phase. The video quality was poor and this was a major concern. Thorough investigation revealed that these errors were due to complexities related to the combination of software, firmware and hardware.

9.3.2   Phase 2 – Redesign and Pre-commercialization

In November 2017, the project steering group faced a difficult choice, either to terminate the project or make extensive changes to the existing Blink design. A decision was made to continue the project, but with strong restrictions on cost and careful management of risk.

This decision resulted in some important changes compared to the first version of the Blink solution. They are listed below:

- Only standard hardware should be utilized. This dramatically reduced development cost and risk and opened up for the use new hardware. This provided the opportunity to exploit Moore's law (Moore, 1965) stating that the cost of Central Processing Unit (CPU) power is reduced by 50 per cent every two years.
- There should be no compromise on user-friendliness as a design principle. This has resulted in reuse of tried and tested graphical user interfaces.
- The intention to reuse the patented, user-friendly remote control if applicable.
- The software should be based on a new Internet-friendly video communication protocol, which resulted in a migration from Session Initiation Protocol (SIP)-protocol to WebRTC-protocol.
- The removal of some additional features including calendar and simplified mail functionality.
- The focus now should be on a stable, secure and user-friendly video call service.
- The new solution should use encryption as well as two-factor authentication and be GDPR compliant.

The new Blink solution is still under development, and the plan going forward is to finalize the commercial test version during Q4-2018, with user testing starting at the end of Q1-2019. A primary objective is to replicate the solution in some of the follower cities as defined by the Horizon 2020 project Triangulum (Remmele, 2018) that focus on smart city solutions. We believe that the Blink solution has the potential to address the 'Smart with a heart' approach (the tag line at Nordic Edge Expo, 2018) and to contribute to traffic reduction.

9.4   THE BLINK PROJECT IN THE CONTEXT OF RESPONSIBLE INNOVATION – A RETROSPECTIVE ANALYSIS

The term 'responsible innovation' can easily be applied outside the boundaries of academia, but perhaps the concept suffers from lack of awareness among industry practitioners. The four underlying principles of responsible innovation can be difficult to comprehend at first glance. Clarifying discussions with some of our academic colleagues in the 'Innovate or Die project' have enabled deeper understanding of the concept. This has supported our analysis of the Blink project in light of the four pillars of responsible innovation: anticipation, reflexivity, inclusivity and responsiveness.

The reader should bear in mind that the authors reside in an industrial rather than academic setting. For example, 'responsiveness' in industry has more than one interpretation. It describes a product's or service's ability to respond to feedback from user tests. It also encompasses feedback from industrial and commercial challenges that need to be addressed if the product or service is to become a reality. The authors argue that the term 'innovation' can only be applied if the product/service represents something new and has a significant and profitable uptake in the market.

9.4.1   Anticipation and Blink

In this section we examine if the project team has considered the consequences of introducing the innovation in the market. Typical questions are:

* What are the consequences for society of the product introduction?
* Can the introduction cause any unintended results?
* Can the product result in lock-in mechanisms, or can the design remove existing lock-in mechanisms?
* Who must and who should be included in the design process of the product?

Regarding anticipation, the Blink project did not start from scratch. First, much of this project was built upon the experiences already acquired from the FUIV project. The use of Full HD video communication in different user groups had been studied, including users with mobility and cognitive impairment, as well as hearing impairment and healthy elderly persons. It was already known that users benefitted from such a product and service and thus it very soon became popular. Being able to use the service in other contexts like friends and family strengthen the use and value of the service.

The FUIV project also gave valuable understanding of unintended use of the service, like how deaf people could create completely new use cases that had not been considered. An example of this is celebrating Christmas using sign language and lip reading. Another important discovery was that elderly users struggled with navigating the different inputs on the TV set.

Second, it was necessary to strengthen the team with hardware, software and health competence. Hence, open innovation was the answer, and the project team consisting of personnel and competence from Lyse, Westcontrol and Norsk Telemedisin was created.

The next question to ask is; what about lock-in? Were we aware of this aspect? The answer is a definite yes. For a very long time we had seen the existing suppliers of video solutions created what we referred to as 'video islands'. In practice, this meant that calls could not be made from one vendor's equipment to another vendor's equipment. This was a major barrier to be able to introduce 'Video For All'. The reason being that despite using the SIP protocol for video communication, each vendor introduced their own twist to it, making interoperability impossible.

Actually, 'Video For All' was the first name used for the Blink product, and it is still a product promise. One of the first decisions was to use the basic SIP standard, and not introduce our own twist to it. The second choice was to seek cooperation with a new company that made this interoperability possible using software to do the 'bridging' between the different islands. This way, the Blink product could communicate between the TV set in the home and a professional video solution in the hospital or care centre.

If the Blink product and service succeeded, this would enable lock-in removal.

Regarding who should be included in the design process, the FUIV-project set out the preferred model. From the very beginning a company with a solid competence in the care segment (Norwegian Telemedicine) was included as one of the project partners.

In the case of Blink, users are not one homogeneous group. The primary user was the person in need of care, but also care givers, volunteer organizations, health professionals and next of kin were users.

At the very start of the project, user-friendliness was set up as the main design criteria for the Blink product and service. For a more thorough description of how user-friendliness and user involvement was the core of our design work, see sub-section 9.4.3 on Inclusivity.

9.4.2   Reflexivity and Blink

Reflexivity challenges innovators to reflect on their choice of business model and evaluate whether its scope is appropriate. The FUIV project provided useful knowledge, especially from the patient/elderly user perspective. Municipalities with experience of using video combined with sensor solutions shared their knowledge with the Blink project team. This ensured that focus was on the care context and the development of a product and service targeting this market.

Phase one

In an attempt to meet the needs of a diverse set of users, the project team aimed to develop the ultimate user-friendly video solution. This was a noble target that became a driver for an ever-expanding list of desired functionalities. The end result was a comprehensive list of design, software and hardware criteria.

Following development of beta versions, user testing commenced. The graphical user interface was tested as a prototype in combination with the technical solution. It was decided that the first hardware products should be tested in the homes of the project team. In addition, the beta version was tested in a healthcare test centre in Stavanger municipality and with a few users in the Horizon 2020 project Triangulum. A decision was made to not include patients in the first beta testing, to avoid unnecessary stress for them.

The test results were overall not very promising. On a positive note, the user-friendly design was well received. The tests provided insights on how to improve the user interface in line with universal design requirements. However, challenges connected with video quality, lip sync and stability issues were reasons for concern. Furthermore, these challenges seemed to be interconnected and often fixing one bug in the software would result in new bugs. The conclusion was that the software stack was too tightly integrated. Some of the hardware architecture choices resulted in bottlenecks, especially when different solutions had to be interconnected.

We will not elaborate further on details regarding the technical challenges. The conclusion was that it was not an option to launch a solution with stability issues, especially not in a market where many of the users typically had cognitive challenges already.

Phase two

The stability challenges forced the steering and project groups to make a difficult choice: terminate the project or make a final attempt to succeed. In order to continue the innovation project, capital and critical video software competency were needed.

An initial assumption had been that functionality requirements added during the design process could only be fulfilled through hardware development. Hardware development represented huge technical, time and economical risk. The steering group unambiguously stated in October 2017 that economic and technical risk must be reduced. This message resulted in the following radical actions:

- Hardware development was abandoned. Thus, the service became hardware agnostic. As a result, some of the functionality in the proprietary hardware was no longer available.
- Open source versions of video codec technology could no longer be trusted or relied upon. The reason being that this part of the software created many of the stability issues.
- Without a stable video platform, there would be no Blink. In order to gain more control over the video platform OMT Technologies was included as a project partner. OMT Technologies' company had competencies in video software development. That would help deliver a stable video platform.
- Moving from 'telecom video standard' called SIP to the newer 'Internet video standard' called WebRTC was a major change. This required completely new software and none of the previous software could be kept.

In retrospect, it is easy to see that these changes were necessary, although reducing functionality was a difficult decision to make. Taking control over the core of the service (video/audio), introduced new options and potential markets beyond the care segment. It is an advantage to deploy and facilitate the video solution across different devices independently of third parties.

The agnostic hardware approach has three main advantages: the benefit of Moore's law (Moore, 1965), increased freedom in choice of hardware and the technological readiness of the hardware.

9.4.3   Inclusivity and Blink

Following the decision that user-friendliness should be the main design criteria it became necessary to strengthen the team with service design thinking competency. Two service designers from Halogen, a Stavanger-based design agency, were hired. Their task was to ensure that graphical user interfaces, interaction design and industrial design of the physical equipment was user-friendly.

Pain points experienced by primary users (elderly people) were well understood. The needs of care workers and health professionals were not well understood. Therefore, a series of workshops were arranged where, for instance, home nurses from different municipalities provided feedback during the design process.

Here is an example of how new knowledge was acquired during the design workshops:

In the first workshop, an older and highly experienced home nurse suggested a new use-case for the product. The municipality she belongs to covers a large geographical area that is sparsely populated. Home nurses (often two during winter months) can spend up to two hours travelling each way to reach remote patients. This is tiresome work especially for older nurses. As a result, younger, less experienced home nurses end up serving the most remote locations in the municipality. The experienced home nurse immediately saw a new use-case (and exclaimed): 'This solution means that I can contribute with my long experience, advising younger and less experienced home nurses at remote locations, whilst I remain in the home nurse centre.' This was a new use-case, particularly relevant for nurses in rural areas.

In 2015, the Innovation team in Lyse was strengthened with an industrial/ service designer. The designer has continued to improve the Blink solution. The designer's work includes developing user graphics to better meet the needs of users with colour blindness. The designer has extended the interaction design to cover the use of touch panels in tablets and smartphones during phase two of the project.

There are different kinds of colour-blindness, the most common form is having trouble distinguishing between red and green. When designing for accessibility and inclusivity, it is important to make sure that colour is not the only method for conveying important information. The Blink remote control uses green LED for arrows, blue LED for enter/ok, and red LED for exit, the same colours are portrayed on the screen.

The first version of Blink was designed around the TV and the remote-control. The new Blink solution included the use of smartphones and tablets, which introduced touch technology. Interaction with smart devices is fundamentally different from using a remote control, so new interaction design needed to be developed. Figure 9.1 shows the design on a smartphone, and how the test tool for colour blindness was used. Figure 9.1 shows a black and white version of the original coloured graphical user interface. The colours in the design were tested for hue and contrast to suit user needs.

User-friendly design features embedded in the Blink product and service relevant to inclusivity are:

- Tactile design of the remote control (that may also be used in the new Blink solution);
- LED-light guidance on the buttons of the remote control;
- Easy to understand user interfaces on different devices like TV, tablets, browsers and smartphones;
- Interfaces tested for colour blindness and contrast for the visually impaired;
- Utilizing the largest screen in the household – the TV set;
- Multi-device interoperability via third party and support; and
- Battery changes avoided through inductive charging. Changing batteries was a challenging task for people with weak tactile skills.

Figure 9.1   Blink smartphone user interface design

9.4.4   Responsiveness and Blink

Responsiveness can be twofold. First, it can be responsiveness to end-user feedback during testing of early versions of a product or service. Second, responsiveness from a business perspective includes how we react to business and technical challenges such as those related to the trade-off between functionality and production cost. The greatest challenge is often balancing user needs with what is viable to develop and sell in a market.

Responsiveness to user feedback
Prior to the Blink project, the FUIV project had provided insightful user feedback following Full HD video testing. The FUIV project gathered input from users with different health and care challenges. This knowledge was applied in the design of the Blink project. It could be argued that the Blink product was borne from the need to make video calls by patients and health workers. If a person with cognitive impairment can use the service, it is likely that almost anyone will find the service easy to use.

During early prototype testing, users expressed an interest in additional functionality that could help develop the Blink product further. Examples include an app that could give patients reminders for taking their medicine, calendar functionality, video for multiple parties and more. As time went by the number of new service features increased.

Increasing functionality requirements led to increased complexity, especially in the software stack. On reflection, this was a costly and time-consuming path to follow.

Responsiveness to end user feedback is crucial if you wish to succeed with your innovation. Although, by adding too many wishes an innovation project can be derailed. One of the important learnings from this project is to focus on the core service and to avoid complexity and additional functionality until the core service is stable and tested amongst users.

Responsiveness to technical and business requirements
An innovation project usually starts with a design process. As the project progresses decisions will be made regarding hardware, technical protocols and target markets. In the business plan, the numbers need to add up. Whilst seemingly simple, all of these early stage milestones represent complex and difficult decisions.

A recent example of responsiveness in the design process came in June 2018. During a workshop, it became clear that the adoption of video-based solutions by doctors faced a security hurdle. In Norway, doctors generate 70 per cent of their income through state refunds. To claim a refund the doctor must authenticate the identity of a patient using an approved two-factor authentication method called certification level four. For this reason, certification level four was included in the new Blink solution. This resulted in a video service with a secure mode targeted at public services such as healthcare. In addition, an open mode was made available to users with less stringent security requirements such as video calls to friends, family and others.

As already described earlier in this chapter, software and hardware technical challenges, high risk in hardware development, led to significant changes in the technical design of the Blink service. In addition to the technical challenges, the target cost especially related to hardware was difficult to achieve.

Without responsiveness to the requirements and challenges outlined earlier, the project would have been terminated before reaching the market.

9.5   CONCLUSION AND FINAL REMARKS

As already mentioned, the definition of the term 'innovation' means that the product or service has reached a considerable market uptake. This is not yet the case for the Blink product since it is work in progress. The authors believe many of the Blink project experiences are relevant for start-ups, established companies and academia.

There appears to be some divergence in how academia and industry interpret responsible innovation and the terminology used. For instance, industry use the terms 'design thinking' and 'service design' while academia uses the term 'inclusiveness'. Industry use terms like 'business plan' and 'SWOT analysis' to cover part of the 'anticipation' term. Another example is the academic term 'responsiveness', where industry sometimes would use 'user-centric innovation' and 'design sprints'.

Below are important lessons learned from this project so far:

- Never underestimate the challenges related to hardware design. This is a constant battle against cost reduction and the risk of becoming obsolete.
- Covering too many 'nice to have features' alongside the core functionality is not recommended. Start with a Minimum Viable Product (MVP) focusing on the core service and test it as soon as possible.
- Always involve end-users. In the case of Blink, there was a heterogeneous user group.

Hopefully, lessons learned combined with responsiveness will result in true innovation and market uptake.

BIBLIOGRAPHY

Moore, G. E. (2006). Cramming more components onto integrated circuits. Reprinted from *Electronics*, volume 38, number 8, 19 April 1965, pp. 114 ff. *IEEE Solid-State Circuits Society Newsletter*, 11(3), 33–35.

Remmele, B. (2018). Triangulum | The three point project Triangulum is one of currently nine European Smart Cities and Communities Lighthouse Projects, set to demonstrate, disseminate and replicate solutions and frameworks for Europe's future smart cites. Retrieved 5 November 2018 from http://www.triangulum-project.eu/.

The largest Smart City Arena in the Nordics! (n.d.). Retrieved 5 November 2018 from https://www.nordicedgeexpo.org/.

10. Hitting the institutional wall – the journeys of three firms from idea to market

Elin M. Oftedal and Lene Foss

10.1 INTRODUCTION

Responsible Innovation (RI) is expected to take society further in terms of purpose, process and outcome. While there is a good understanding of how public-sector priorities and values may drive responsibility, not much is known of how small start-ups navigate in these waters. Further, research has shown that many patients and their dependents develop innovative products themselves, however, less is known of the challenges that meet innovative users of the health sector. Being an entrepreneur in the healthcare sector is associated with different challenges than other sectors, as the public sector is a powerful constituent. An important question is what kind of institutional challenges these entrepreneurs are facing, when commercializing products based on the principles of RI in an emerging part of a sector.

We suggest that institutional perspectives offer a useful theoretical lens on how innovation may emerge despite the tensions between different norms and practices (De Vries, Bekkers and Tummers, 2015). Interestingly, the literature on institutional change does not focus explicitly on innovation and technological change, nor does it explicitly take into account relations between other forms of system-building activities and institutional entrepreneurship or institutional work. Therefore, building on regulative, normative and cultural-cognitive pillars (Scott, 2014) this chapter invites the reader to become familiar with the institutional forces that shape and constrain entrepreneurial encounters. Being inspired by metaphors such as theory-constitutive (Boyd, 1979) we suggest 'institutional wall' as a powerful image of how entrepreneurs in new e-health start-ups face a structure built for other purposes than easy commercialization of responsible innovations in the Norwegian health sector.

This chapter takes the empirical context of entrepreneurs in the emerging e-health sector related to primary healthcare services in Norway, as a point of departure. Not only does the health sector continue to represent a significant proportion of GDP in most economies, it is also central to the resolution of many pressing and often 'wicked' social and environmental problems (Rittel and Webber, 1973; Weber and Khademian, 2008). Structural and related cultural characteristics can both stimulate and constrain innovation (Potts and Kastelle, 2010; Borins, 2001; Vickers et al. 2017).

Vickers et al. (2017) argue that asymmetric incentives are where unsuccessful innovations are punished more severely than are successful ones rewarded, and there is also adverse selection by innovative individuals against careers in public services. Earlier, Bate (2000), found that clinicians within a hospital work environment were resistant to innovation due to 'conservatism' and 'defensive cultures'. We therefore ask the question: How do entrepreneurs presenting responsible innovations in e-health experience their institutional context?

This study contributes to enlighten the application of Scott's (2014) institutional pillars by suggesting that the regulatory, normative and cultural cognitive forces make an 'institutional wall'. We integrate knowledge from methodological literature in management and strategy (Alvesson and Sandberg, 2013; Gioia, Corley and Hamilton, 2012; Alvesson and Kärreman, 2007; Hunt and Menon, 1995; Boyd, 1979) in suggesting this new metaphor ('institutional wall'). Our results exemplify how commercialization of e-health services, in a well-organized industry like the health sector, are challenged using new technologies. Our findings illustrate how three entrepreneurial firms are embedded in an 'institutional wall' challenged by an infrastructure that is not conducive to responsible innovation.

The rest of the chapter is organized as follows. The next section describes what we mean by 'institutional wall' using institutional theory (Giddens, 1984; North, 1990; Scott, 2014) and explain why such a 'wall' makes responsible innovations challenging. This is followed by section 10.3, which describes the Norwegian health sector. Section 10.4 reveals our methodology followed by the empirical findings in section 10.5. Section 10.6 analyses and discusses the findings. The final section outlines implications for theory and policy and concludes the chapter.

10.2 THE INSTITUTIONAL WALL: REGULATIONS, NORMS AND CULTURE/COGNITION

The term 'responsible' evokes concepts and actions that involve actors from different institutions (entrepreneur, patient, public healthcare, private healthcare, financial institutions and so on). Consequently, an institutional perspective may offer new insights and contribute to increased knowledge of the context in which RI is pursued within the e-health sector, and activities take place in its institutional context. Institutions has been defined as 'the rules of the game (North, 1990) and serves as constraints or enablers of certain behaviours. Thus, institutions reinforce behaviour that strengthens or maintains the existing institutional structures (Giddens, 1984; North, 1990, Scott, 2014, Palthe, 2014) – but work against behaviours that are associated with contradicting the existing. Further, institutions can be defined as regulative, normative and cultural-cognitive elements that individually or collectively work to create shared meanings and frameworks that support social order (Scott, 2014). Consequently, institutional theory informs our research question by identifying how institutions facilitate or hamper entrepreneurial efforts by start-ups in e-health. 'The wall' refers to an institutional framework that is constraining start-up activity so much that it is detrimental for their survival. We look at the three dimensions of the institutional framework as identified by Scott (2014): *The regulative pillar* concerns mandated specifications including laws, governance and monitoring systems. To be legitimate within the regulative pillar, the formal laws and regulations have to be followed as these are the building blocks of this dimension. Any change in this pillar also follows a legal obligation, however sustaining the pillar requires respect and even fear of breaking the laws and rules. Actors within this pillar act a certain way because they have to (Delbridge & Edwards, 2013; Heracleous & Barrett, 2001; Huy, 2001).

Applied on the health sector, start-up behaviour that goes against the existing regulative system, will be challenged. The Norwegian healthcare system is heavily regulated and any start-up activity must be attentive of that.

The *normative pillar* incorporates values, expectations, including roles, repertoires of actions, conventions and standards. This pillar deals with the informal, unwritten 'rules of the game', for example, what actors within the framework consider as wrong or right behaviour (Meyer and Rowan, 1977; Suchman, 1995). Socio-political legitimacy, on the other hand, does imply the act of normative evaluation. In a socio-political legitimacy judgment the observed features of an organization, its structural attributes, and outcomes of its activity are benchmarked against the prevailing social norms: the actor renders a judgment as to whether the organization, its form, its processes, its structural outcomes, or its other features are socially acceptable and, hence, should be encouraged (or at least tolerated) or are unacceptable such that the organization should be sanctioned, dismantled, or forced to change the way it operates (Aldrich and Fiol, 1994; Kostova and Zaheer, 1999; Meyer and Rowan, 1977; Scott and Meyer, 1991; Suchman, 1995; Bitektine, 2011; Palthe 2014; Oftedal, Iakovleva and Foss, 2017). Legitimacy here rests on actors' morals and the norms of the society. System change stems from a sense of moral obligation to the society, while sustaining the pillar rests on moral obligation. Actors here base their behaviour on an 'ought to' mentality. Applied in

Table 10.1   *Regulative, normative, and cognitive elements associated with behaviour*

	Regulative	Normative	Cognitive
Legitimacy	Legal systems	Moral and ethical systems	Cultural knowledge and interpretations
System Building blocks	Policies and rules	Work roles, habits and norms	Interpretation, beliefs and assumptions, values
System Change Drivers	Legal obligation	Moral obligation	Change in interpretation of knowledge
System Change Sustainers	Fear and coercion	Duty and responsibility	Social identity and personal desire
Behavioural foundation	Have to	Ought to	Want to

Source: Palthe (2014), s. 61.

our case, these are the mechanisms that the entrepreneur faces when trying to create innovative patient-friendly solutions.

Finally, the *cultural-cognitive* pillar encompasses predispositions and symbolic models for individual behaviour regarding the acceptance of entrepreneurship. This pillar deals with the knowledge that resides within the institutional framework and how actors interpret existing and new knowledge. Legitimacy here is based on the knowledge of the society. This knowledge is filtered and interpreted through a cultural lens. The building blocks of this pillar are the beliefs and assumptions based on actor's values. System change is based on a change in knowledge or interpretation of knowledge. Legitimacy thus implies different 'modes of action'. Cognitive legitimacy spares the organization from increased scrutiny and distrust of external social actors by making the organization understandable and taken for granted for its audiences and permitting cognitive typification of this organization into a pre-existing category (Berger and Luckmann, 1966; Meyer and Rowan, 1977; Suchman, 1995, Palthe 2014, Oftedal, Iakovleva and Foss, 2017). Sustaining this pillar is based on social identity linking to the established convention of knowledge. This means that innovations will be understood through the filter of the existing knowledge. In the case of cognitive legitimacy, the evaluation stops when the organization is classified as a member of some already known and already legitimate class of organizations (Barron, 1998; Bitektine, 2011). Table 10.1 follows Palthe (2014) and presents the underlying logic of the three dimensions, and how they differ.

In suggesting the metaphor of the institutional wall, we apply Scott's institutional pillars in order to study how RI entrepreneurs manage their institutional context in providing a new service for patients outside the public healthcare system. By managing or handling the institutional context we aim to understand how the pillars facilitate or hamper entrepreneurial efforts.

10.3   THE INSTITUTIONAL CONTEXT: CHARACTERISTICS OF THE NORWEGIAN HEALTH SECTOR

The Norwegian example is intriguing, since it is almost exclusively publicly funded through taxation, and most hospitals are publicly owned and managed. The system includes a fairly strong primary care sector with family physicians to various degrees acting as gatekeepers to specialist services. Furthermore, this system includes devoting special consideration to the needs of those who have less chance than others of making their voices heard or exercising their rights. Issues of limited access are now, however, challenging the thinking about a healthcare system based on solidarity (Holm, Liss and Norheim, 1999). Norway bases its model on the Beveridge model where healthcare is viewed as a human right and a citizen's privilege.

We claim that the flip side of a government-controlled model with a high focus on equal access and primary care, may be that it is a top-down system with few incentives for patient involvement. Further, the imperative to hold down costs may imply that the newest technologies are likely not easily available. The buying process is characterized by buyers who are often governed by the public sector, or face strict regulation. Consequently, the market consists of a hard-to-reach and diverse audience of buyers. Since healthcare is a public right in Norway, the private market is underdeveloped, creating a gap between the user (patient/dependent/health worker) and the customer (public buying systems). Therefore, digital healthcare may appear to have high promise for patient-centred high-quality healthcare delivery at an affordable cost and open to all citizens. The newly established 'Directory for E-Health' launched the 'National Strategy for E-health – 2017–2022'. They point towards six strategic areas: digitalization of work-processes, better management of patient flow, better use of health data, new methods of health aid, a common foundation for digital services and national governance of e-health and increased execution within the strategic areas. Although digital delivery may seem rational and provide an effective implementation, there are clearly challenges. While start-ups have been engaged in RI, there is a risk that the established actors are favoured as a trajectory emerges. The entrepreneurs' actions are embedded in an emerging institutional setting, where rules, norms and belief systems are yet not set.

Initially, we described the institutional structure as consisting of a regulative, cognitive and normative dimension. Figure 10.1 provides an overview of the Norwegian healthcare system including financial links, patient flows and the GP referral system. The original model (Mørland, Ringard and Røttingen, 2010) showed that the central government allocates resources for the regional health enterprises which again are responsible for health enterprises such as hospitals. The patient is in direct contact with the system through the primary care giver (i.e., the GP) or the municipalities (i.e., nursing homes). The primary care giver will be the gatekeeper for the hospitals. The challenge with the model (Mørland, Ringard and Røttingen, 2010) is that it is omitting firms and private healthcare providers. Figure 10.1 illustrates how private actors may help to close the gap between the public sector and the user through digital technology. It also shows that there is an institutional framework around the private actors, such as regional authorities, governmental agencies to promote innovation, in addition to knowledge institutions such as universities and the Norwegian Research Council.

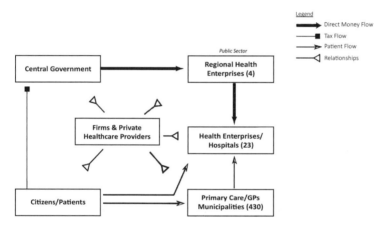

Figure 10.1   Overview of the Norwegian healthcare system

Source: Adapted from Mørland, Ringard and Røttingen (2010, p. 399).

10.4   METHODOLOGY

Based on our exploratory research question we have chosen a methodology suitable for getting in-depth insights from entrepreneurs' challenges in commercializing digital e-health solutions in a responsive way. This study is a part of a larger project where both start-ups and established companies are monitored over time to understand how they work in terms of RI. We therefore had interview data from several start-up firms. To back up the data from these companies, we had interview data with stakeholders, discussion on online fora and newspaper articles.

For this particular study, we wanted to inform the reader of some of these cases and thus we pulled out early stage firms from the larger sample to present a narrative approach. These were all companies that had applied one or more stages in the RI framework, such as purpose and process. The companies have an idealistic vision of improving their users' lives (purpose), further they have applied inclusion, anticipation, reflexivity and response in some manner (process). In order to apply the three pillars of influence (Scott, 2014) we wanted start-ups which had been influenced by their entrepreneurs in an early phase, before they had grown older, become immersed in the market, and developed more rigid business routines. Consequently, we chose three start-ups with a very active entrepreneur and which had a compelling story in relation to the institutional wall. These companies were all started by former and current users of the health sector, typical 'innovative users'. We also chose firms that had different product segments which enabled us to explore different parts of the institutional environment. We decided to label them Voco, Cora and Medicus which are Latin for their core activity.

We follow Gioia, Corley and Hamilton (2012, p. 25) in that qualitative research can and should be able to stand on its own. In our company-interviews, we used an interview guide, where the RI questions were based on Stilgoe, Owen and Macnaghten (2013). Further, we included questions related to the nature of their innovation, their business, their stakeholders and their entre-preneurial journey. The interviewees identified different segments of their institutional structures that were important stakeholders for them, namely, governmental innovation agencies such as Innovation Norway, Norwegian Research Council, universities, research institutes, investors, insurance com-panies, hospitals, local governments and politicians. The interviews followed an ethno methodologist approach where we looked for the processes by which the interviewees made sense of their interactions and the institutions through which they operate (cf. Feldman, 1995). We were inspired by Alvesson and Sandberg (2013, p.146):

> Crucial here is to challenge the value of an established theory or a framework, and to explore its weaknesses and problems to the phenomenon it is supposed to explicate. It means generally open up, and to point out the need and possible directions for rethinking and developing it. We consequently suggest a methodology for theory development through encounters between theoretical assumptions and empirical impressions that involve breakdowns. It is the unanticipated and the unexpected – the things that puzzle the researcher – that are of particular interest in the encounter.

In practice this meant that when we sensed when the interviewees became engaged, we let them talk and probed with following up questions in order to understand more the nature of the institutional complexity they were embed-ded in, and how this affected them. As Scott's institutional framework is well applied in prior research, our aim is first and foremost to learn about the unanticipated experience entrepreneurs have when facing the institutional complexity in the health sector. Thus, we view our data as an inspiration for a critical dialogue between Scott's institutional pillars, the RRI framework and our empirical work (cf. Alvesson and Kärreman, 2007). Finally, we do not claim generalizability of this study. This narrative account of three entrepreneurial cases is meant to stimulate a more theoretical generalization where the metaphor of institutional wall could embrace the vocabulary in institutional theory.

10.5   FINDINGS: THE ROLE OF RI IN FOUNDERS' MOTIVATION FOR THE START-UP

We start this section by introducing a description of the start-ups and the entrepreneurs' own motivation for the company and how it relates to RI.

10.5.1   Voco

Voco was established in 2008. The entrepreneur (a computer scientist) experienced a severe motorcycle accident and during a prolonged hospital stay he observed how the healthcare personnel were interrupted during their workday and how this affected him as a patient. Sensing the healthcare personnel work situation and the need for an interruption management system within the hospital environment, the entrepreneur took on the challenge to develop a context-aware interruption management system. This system automatically monitors availability on mobile devices detected from sensors that receive information based on the situation or location (context) healthcare personnel are in, and then interruptions can be avoided, and calls can be redirected to other available resources.

The system is designed to gather resources and allocate resources, such as medical doctors and nurses, intelligently. That is, avoiding disturbances, but at the same time get the resource needed at the time it is needed. The ultimate impact on this system would be to install it in all hospitals in Norway.

> People and patients are the focus. Everything else just has to work. You have to spend some time introducing a system. That's what has been the failure over the years; pushing on systems that have not been ready, and have not worked the way the user wanted, needed or expected. The user has often not been fully involved in the development of the systems. Often just the leaders are involved. If they say it's user involvement, then it's usually the leaders … I had to understand the users and that the system is based on the users. I saw myself not as "the creator", but rather "the intermediary" who translates their need into an IT program. One of the challenges is that there has been too many failures in using the new technology. By

introducing systems that do not fit or work properly, people get tired easily. One has to be hundred per cent sure that it works.

The quote shows Voco's dedication to the user. We view the founder of Voco as having a clear purpose to help the user and who includes his user in the development. This is evidence of RI and Voco has built a system which is perceived well by users. We wanted to follow Voco's journey from idea to market, indeed we had data from earlier projects with this company. However, it became clear early that Voco was distressed by several obstacles that he met. Just two years after our project start, the entrepreneur shelved the company and took out a sick leave.

10.5.2   Cora

Cora is a technology-based company, started in 2014 based on product development that was executed in a previous firm. The company is working with sensor development and advanced signal processing techniques, exploiting multi-sensor data fusion to accurately describe biological processes mathematically. Cora incorporates a diverse mathematical approach into signals processing techniques that enable a high precision assessment of heart performance and detection of continuously abnormal heart rhythms in real time. Cora's product makes monitoring of chronic heart conditions in children in an easier and more exciting way for both children themselves and their parents/ caretakers. The entrepreneurs' expressed goal is to empower individuals with chronic health conditions involved in daily self-care. This is based on the fact that 90 per cent of follow up on chronic health conditions happens outside hospitals, where patients do not have access to advanced medical technology.

> I suffer from the condition myself. I'm basically a part of the market. The first families we met, were so much willing to solve this problem. They have experienced that the doctors will not help them, so they share, help and contribute in almost any way. We had a family who want to share because the doctors don't hear them, neither the healthcare-system; we want to get this message through. Everyone seem to be looking in to this from the perspective of the healthcare system, but not so much from the patient's perspective.
>
>   For us it was very important to get this support from the patients and we've got it. The patients have changed our view on their needs. I would say they made it broader – they opened up for much more new information. To begin with, we thought of just looking in to the pulse frequency, but then many parents asked us, and then doctors confirmed about this oxygenation. Level of oxygen in the blood has to say a lot about general heart condition.

Cora had a very dedicated co-founder who was an example of an innovative patient and a user of the technology. They also fell into the category of RI as they had a purpose to help existing patients and their dependents. They included users in their development and reflected upon their own practice. However, they had several obstacles. They wasted time collaborating with the university and the university incubator. Further, they did not get the needed investment. They finally exited, after some years of struggling.

10.5.3 Medicus

Medicus is a small, but rapidly growing digital health start-up. The founder of the company had worked in finance for almost a decade and experienced the shift from traditional banking services in online banking and trading where financial data was available on a single platform. He discovered that these solutions were not developed in the health sector. Based on his own experience with a lack of efficiency in the healthcare sector, their aim is to change the face of Norwegian healthcare by offering a digital platform to communicate and share data with healthcare professionals. As a patient he felt the urge to transform health data into something that is owned and managed by a patient. The first step in the company was to develop a mobile app allowing video conferences with doctors. Their first major feature is video consultations with doctors on apps for the web and smartphones. In their view, the current healthcare system is unsustainable and inefficient and they are therefore working on solutions that deliver healthcare through a digital platform available on consumer electronics. Healthcare is different from many other industries and requires us to deliver secure systems and navigate a strict regulatory legal environment. Backed by a private clinic, they have access to an existing medical system to build upon. While ordinary doctors' appointments in Norway costs NOK 220, they can offer it for 350.

> I just do not think it's fun to wait. Standing in line is one of the worst things I know, and it seemed very corny to go down to the doctor, wait for 40 minutes and then go into the office and talk to him for seven minutes and then go out again. I think it just seemed very little effective.
>
> Use of people's time and the premises they have, and also my own time. It was the starting point really., I think the digitalization in general and health is incredibly exciting. It is so incredibly far behind other areas of society as all other consumer-oriented services have come further. The background I had from finance has been as conservative in many ways and is as concerned with security and privacy as maybe health. I knew the solutions that health needed existed. We already had both business models and security systems. We had all the tools ready in that industry. Basically, when we developed the application, we were somewhat involved with potential customers. But it was only when we got used to the application that we built dialogue with customers for testing.

Medicus also has a clear purpose and is very focused on the user. They participated in all activities of RI and responded to the users they included. The difference between Medicus and the other companies is that Medicus is more consumer oriented and work with the private rather than the public sector, although they have to cooperate with the public sector. The company has worked well with investors and stakeholders and has achieved growth in income and employees, since they are in a growth phase, they have had negative equity and profit.

10.6   THE INSTITUTIONAL WALL

We describe the 'institutional wall' as an institutional structure that may enable or constrain entrepreneurial actions. Recently there has been substantial effort in e-health; the 2017–2022 strategy includes new technological solutions to improve access to health data, to digital services and to an improved quality of healthcare. This should lay a good foundation for innovative start-ups, however, the Norwegian health system has been designed within a strong public sector, with the aim of protecting patients, but also of controlling costs. This places some constraints on the actors in this market. The three starts-ups, Voco, Cora and Medicus, had several challenges with their surrounding structures and gave voice to some of these concerns. Here, we illustrate issues they are mostly concerned and passionate about, before discussing them in light of the theoretical framework. We organize them within the framework of the regulative, normative and cognitive dimensions.

10.6.1   The Regulative Dimension: A Major Challenge With Careful Positive Change

The regulative dimension refers to formal regulations and legal systems as the bases of legitimacy (Table 10.1). Here we present the entrepreneurs' experiences with the existing framework.

Voco
There is high national control because you have to be accepted by the health regions. If you're going to do something in the regional hospital, the whole health region must approve of it. And if you raise it to a national level, it must be approved by the National ICT. So that the decision-makers in the healthcare industry in Norway are on such a high level with large organizations and you must be approved for everything. If you're not big enough to pull in the effort you want, then it's almost completely impossible to get into that world. It's a world that only works for the big players. They have blocked it in and they have made rules that indicate that the bids are also big, so you do not have the chance to come in as a small player without having one of the big ones in the back. You do not have the opportunity.

Medicus
Another thing is that the procurement and the decision-making in the municipalities and hospitals is very difficult. I think there are 30–40 companies that tried to work with the municipality in some aspect of the health area that went bankrupt. But there is a development here. A new framework called 'innovation – partnership' is approved. Through this, the hospitals can buy from smaller innovative suppliers and where there is less strict requirements to the procurement/tender process. This is a good development and gives easier access for small firm[s].

These quotes show how the entrepreneurs contemplate the regulative aspects of their sector. It is rigid and large, however, according to Medicus, it's moving in the right way.

Further, Medicus is expressing contentment about aspects of regulation in the sector. The norm for an information security function is a guiding set for the companies:

We have a 'Norm for information security' in health and care services, which is the standard for safety and privacy in Norway. This is based on a unified set of requirements in the legislation. It's a specification of demands, where several laws have been built in such as the 'Health Personnel Act' and the 'Privacy Act'. We have to sign and that's incredibly useful, because we know how to handle security issues. We are lucky in Norway where such a framework has been developed. It makes it easier to comply with the laws.

The regulative dimension is a strong barrier to entry for the start-ups. Through a carefully regulated framework, larger established firms are prioritized, something that concerned our respondents. However, positive changes in the regulative framework give hope to a careful optimism. It is also obvious that when well formulated, the regulative framework can be an enabler, such as with the 'norm for information security'.

10.6.2   The Normative Dimension: Comradely and Fumbling

Norms specify how processes and activities should be accomplished to be accepted in a society. In the material we have analysed, there are several challenges associated with the normative dimension. From the entrepreneurs of the start-ups we observe a mismatch between obligations and norms from the start-up society towards the institutions. The 'who you know' approach is frustrating to the entrepreneurs because while they work consciously on creating innovative and responsible solutions from the health sector, they sense there is not an apparatus or an understanding in society for what they accomplish. Thus, their relation to the public sector becomes coincidental and they spend a lot of time and resources 'stumbling' forwards, hoping to meet the people that can help them with the next major steps.

Voco

Voco goes very far in critiquing the Norwegian networking style and makes clear hints that this might not be fully ethical and very harmful for start-ups.

> I have the impression of camaraderie: "I know people whom we can pull on for resources and then we support each other, and we pat each other on our backs and we just pull in what we know" … This is the Norwegian form and we acknowledge it's an accepted part of doing business in Norway. I actually think it is a big problem. It is a small society where too many are comrades. And if you do not know the right people or have the right interests, then you're almost unable to do business, you must rely on pure luck.

According to Voco, this form of camaraderie follows the person in dealing with the institutions. There are also other complications Voco faces in regulated healthcare system:

The tools that were in place to help the start-ups also have challenges. One of the tools that the municipalities can use is to start a 'pilot project'. However, Voco's frustration is that these pilot projects do not lead anywhere. Voco calls this 'the pilot disease':

> The municipality is happy to give you a pilot project but when it comes to the actually signing a contract, they do not have money nor opportunity. Nurses or assistants generally like the tool and they enjoyed the pilot. It's a great opportunity for us. While decision-makers have no decision-making ability or no money then, to buy it. That's what they call the "pilot disease".

Cora has been able to access a governmental programme in Finland, which they found more effective and professional in dealing with start-ups. They lament on the time and resources wasted on meeting people without interest or competence in the area in Norway. They discuss the difference between fumbling to get the right resources and compare with the Finnish system.

> It really depends on who you meet. We had a meeting at a university hospital in Finland, were we were immediately directed to a research group working with the heart condition we are targeting. So, we didn't have to explain the problem to them and that there's a need. They knew it before we came. It was a huge change from when we had the first meeting at our local hospital in Norway, where there were a couple of doctors who were not really into this topic. They were working with kids with heart disease, but it might be they didn't work with this exact problem. So, we didn't get that much support here, to begin with.

Cora also turned to the university in order to get assistance to develop some solutions. While the university was positive to help the firm, there were no systems in place help the young start-up. Consequently they lost a substantial amount of time.

> We came to the university and asked: "We have this plan for the development, can you through your department, produce this prototype? The university was like our "middle person" We had to do that otherwise we would not get the money. We signed the contract with the university, and it was supposed to take them two months, but it took them nine months. It took them one month to answer our emails. When I called the university and asked them send us an invoice it took them three months.

On the other hand, Medicus notes that municipalities and hospitals do not trust the industry to develop good solutions for them, but try to come up with home grown solutions:

> I have noticed two issues: Institutions like hospitals and municipalities, often prioritize to solve their own problems before the customer's problems. They base their actions on what is possible for them, rather than what the customer wants. The other issue is that the purchasing possibilities and decision-making capacity in the municipality is very difficult. I have come in contact with 30–40 companies who tried to work towards the municipalities in different health areas that have gone bankrupt. The belief is that the public organizations should build everything themselves. That hinders innovation and prevents companies, like us and others, from delivering good solutions.
>   Voco reflects on how people within the formal systems do not always have the correct background or information to make the right decisions about which technology to use or buy.
> I think it complex to fit with how the hospital, health and hospital world are set up. To be accepted into the health sector is very difficult and you are quite lost also in relation to a sale. It's also the case that you are competing to deliver on the big complex systems which also the larger companies compete for. We experience that it is the wrong people who put their foot down. It is not the actual users of the technology or those that have right technical skills.

Medicus also pointed towards a superficial treatment of entrepreneurs. While the entrepreneur is proclaimed as a hero, a change agent and a pioneer, there is little substance behind the choruses of politicians and academics. He illustrates the pitching competitions as follows:

> Being good on stage is not the same as being a good entrepreneur. Spending a lot of time on pitching competition takes time away from core activities in the business which is product development and sales. I also suspect that the stakeholders, like banks, insurance and oil companies prefer to organize competitions because it's a cool way to show that they help. It's amazingly good branding for them. The ecosystem around the entrepreneurs benefit from the arrangements and have strong

incentives to initiate these projects. It often feels like we are like invited entertainers. Here comes the clowns who will be on stage and entertain the rest. But it's not about us. It's not about making a business. There may be twenty companies, pitching for two minutes each. It's then forty minutes for the entrepreneurs to show off. And then there are hours of speeches from Innovation Norway, the Crown Prince, from politicians. And the investors we need, they're not here, there are very many things you could do that would make it incredibly much more valuable for the entrepreneurs, and not just the ecosystem.

We found that there are several norms that function as obstacles to the development of the start-ups: (1) the waste of time of searching for resources; (2) the dependencies on networks and 'who you know'; (3) the favouring of larger firms; and (4) the lack of efficiency for tools that are in place to help entrepreneurs (i.e., pilot programmes, start-up competitions). But favours the established firms.

10.6.3   Cognitive Dimension: Lack of Belief and Tough Access to Finance

The cognitive perspective in institutional theory stresses the importance of the taken for granted knowledge in the society (Meyer and Rowan, 1977; Scott and Meyer, 1991; Suchman, 1995). New knowledge will, by definition, meet obstacles as it challenges existing knowledge. We also link the cognitive to investment, as investment usually goes towards taken for granted solutions (Bitektine, 2011). Our data reveals that the new knowledge that start-ups are presenting is not readily accepted by key stakeholders. Examples of that is the lack of belief in their technology and a lack of willingness to fund it.

Voco claims that there is a lack of belief in technology from Norwegian companies. The cognitive dimension in Norway seems to embrace foreign technology.

> It is the Norwegian history of technology development. We have always sold the good technology out of the country because companies in Norway have failed to invest. Just look at such simple things as GSM; it was developed at Telenor and was sold out of the country because no one was investing. It appeared too uncertain. The sentiments are that we are too small in Norway, we dare not trust ourselves. So, we're pushing ourselves a little down in Norway. The technology is so much better abroad.

The cognitive dimension of institutions signifies accepted knowledge, both of the product or technology that the company is presenting and of the organization itself. Further, cognitive legitimacy also encompasses the background of the management team and of written documentation about the organization (Barron, 1998; Hannan and Freeman, 1977; Meyer and Rowan, 1977; Scott,

2014; Suchman, 1995, Bitektine, 2011). A substantial issue is the general access to capital, as it is difficult to finance the development from start-up to a successful growth company. Growth requires more capital than the owner and the business can manage to cover. During this phase, public R&D and the entrepreneur's own funds end before the idea or product is close enough to the market to be commercially interesting for seed grain or venture capital (private and/or public equity that goes into an early phase into ownership and innovation and innovation projects). Special venture companies rarely enter before they realize the contours of a business and an expected market launch within a few years. They expect seed grain funds and business angels to enter earlier.

Voco

I have received reasonable support from my organization. At the same time, it's a matter of being a prophet in your own house. And it's not easy within the system either. We had some meetings with Innovation Norway and some investors to assess our potential. We wrote two proposals, and the last time we needed half a year, but we managed to get an interview. However, we fell out of the competition because someone in the committee thought that a larger company should be prioritized. We experienced that the regional governmental agency didn't help our process and we fell out of the competition.

Cora

Everything stops just because we don't have this 1.2 million to cover our expenses. So that's why we are applying, or we have the meeting with investors, with the idea to ask for 1.2 million, to produce prototypes and to get moving. Twentieth of February, I'll have the meeting with one of the possible investors, he's a partner at McKinsey. He believes that early capital in Norway doesn't exist, and it's killing all the research-based start-ups, particularly the ones that are not connected to the government. He was the first person who openly told me that if you want to survive as a research-based start-up in Norway, you have to be a part of a governmental institution, otherwise you don't survive. His idea was he wants to use his capital to cover this. So, he has invested into research-intensive start-ups but without governmental ownership.

Medicus

The largest challenge in Norway is undoubtedly funding. It's so embarrassing to talk to Ukraine, where we have to say: "we do not have money in Norway". We are one of the richest countries in the world! … So, everyone thinks that Innovation Norway (governmental funding agency) is about innovation for entrepreneurs and start-ups, but it's not. Innovation Norway is, if we look at how the money is spent, mainly a distribution organization for regional policy. That's it. Why should you sit as a bureaucrat and evaluate how well our idea is? We have had two meetings with them, spending a lot of time on two applications – which were both rejected!

In summary, the cognitive dimension signifies how a certain knowledge or interpretation is accepted in society. The lack of funds and belief in Norwegian start-ups in e-health, signifies a disconnect between the knowledge that these companies represent and the interpretation of it, from the larger society.

10.7   DISCUSSION

This chapter focused on how three promising Norwegian start-ups demonstrated their responsibility through the innovative product that they have developed and their wish to contribute and solve a challenge of the society. While their technologies are tested successfully among different stakeholders, each of these companies are having issues with the institutional framework.

This analysis show that the regulative, normative and cognitive dimensions are supporting a system where these start-ups are excluded.

The institutional wall, in our examples, is quite dense. *The regulative pillar* is initially a major challenge for start-ups in e-health. The system is set up for larger actors and the prospects for a small new company to enter the marketplace were described as quite bleak. Heavy regulations ensures control for the government, but creates major barriers for the start-ups. However, positive developments were described as new laws and regulative structures are being passed to make it easier to enter the system. The way that these laws are enforced makes it easier for actors to navigate in the system. *The normative pillar* was also problematic as many norms and routines created obstacles for the entrepreneurs. First of all, the coincidental way that entrepreneurs are treated, based on luck and 'who you know' was time consuming, confusing and frustrating for the founder. Further, the attitude that governmental institutions want to create their own solutions for their challenges and not leave this to small companies with special competencies, was frustrating and indicates a lack of trust. Finally, the initiatives that did exist to support the entrepreneurs, such as 'pilot projects' are not very effective. The cognitive dimension is understood as the prevailing knowledge of a certain issue in a certain society. As start-ups often present new knowledge through innovation, they are seen as change agents of a society and will challenge the prevailing knowledge. The lack of investment in the new ventures indicates that the knowledge and understanding of these new companies in the e-health sector is limited.

There is an inherent conflict between institutions and innovation, since the latter promise change and the former wants to preserve change. However, RI should ideally have a better chance of succeeding, since it's hinging its technology on challenges that the society at large agree to tackle. While the start-ups in our example are purposefully trying to help their potential customers, including them and responding to them, there is no advantage for them within their institutional framework. We demonstrate that the role of the institutions could be emphasized more to understand how the start-ups are accommodated or opposed in the institutional framework.

To illustrate our findings, we extracted the model in Figure 10.2 from our earlier Figure 10.1. Based on our cases, we have seen that e-health start-ups may have responsible relationships with their users (patients, caregivers). Still, when commercializing their technology, there is a dense 'wall' between the start-ups and the healthcare system that might be hard to break through. This wall is a way for the government-controlled health sector to keep control over quality, cost and resources of the healthcare system. However, if the wall is so dense that RI start-ups give up, it is not working optimally. We have found evidence through our cases that the institutional wall in Norway might be too dense to ensure and help the development of the creative, local start-ups, and instead prioritizes larger companies.

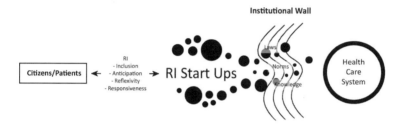

Source: Author's own.

Figure 10.2   Model of institutional challenges

10.8   CONCLUSION AND IMPLICATIONS

By acknowledging the emerging e-health firms within the government-controlled Norwegian health sector, this chapter explored the institutional context of three e-health entrepreneurs who use the principles of responsible innovation in their operations. Theoretically, we acknowledged that e-health entrepreneurs face institutional complexity and defined the institutional context as consisting of regulative, normative and cognitive pillars that together form a 'wall' that entrepreneurs encounter when they are establishing their start-ups.

Our data consisted of three start-ups, drawn from a larger sample of e-health entrepreneurs in Norway. We aimed for a narrative account of these three cases in order to stimulate a more theoretical generalization where the metaphor of institutional wall could embrace the vocabulary in institutional theory. Further, that this narrative account opens up for a more subjective understanding of specific challenges and allows readers to put themselves in the shoes of the entrepreneurs. On the other hand, it functions solely as glances into the realities of the specific founders and thus this chapter does not claim any representativeness of the findings. Our goal is to create awareness of the specific situation some e-health founders are in and inspire new research areas.

Our findings revealed that the institutional wall was considered quite 'dense' by our informants. The regulative pillar is a major challenge for start-ups in e-health. Yet there were some positive elements as new laws and regulations are passed to make it easier for entrepreneurs to enter the system with their new products and services. Actually, the normative pillar came through as harder obstacles, as values and norms in established institutions are working against the entrepreneurs. Finally, the cognitive dimension shows that start-ups in e-health are discouraged by the lack of investment and belief in their technology.

Our conclusion is that e-health entrepreneurs pursuing responsible innovation hit a 'thick' institutional wall which may obstruct and delay them in their development. The practical implication of this is that products of e-health entrepreneurs have limited chances of being thoroughly tried and tested on the market. Since these companies are new and have limited amounts of money, the consequence is that the offered products are either replaced by inferior products, products from established international companies or no products at all. This seems like an inefficient allocation of resources, since local entrepreneurs are putting their knowledge, experience and resources into their projects.

We acknowledge that our study has implications for developing an institutional theory of responsible innovation. Our study should inspire scholars within innovation and organization theory to refine the regulative, normative and cognitive dimensions in the setting of industries that have a mission of responsibility. The research field needs more theoretical advancement on how institutional walls exclude industry development and entrepreneurial opportunities. In conceptualizing the institutional wall in more detail and developing empirical indicators, the field will develop theoretically and improve its clarity.

For practitioners, we advise to reduce the hindrance of the 'wall', by connecting to stakeholders throughout the process. Indeed, the principles of responsible innovation may be useful in connecting to important stakeholders to understand the various challenges. Entrepreneurs in new industries are novices, and have few established and successful entrepreneurs to relate to. Another implication for founders could be to move deliberately and take minimal risk at each step, as institutional arrangements may only change gradually. They can do this, by not taking unnecessary financial risk, and to not lose their initial source of income.

Implications for policy-makers, such as politicians at local, regional and national levels and leaders of public health organizations, would be to carefully consider the situation and develop an ideal future scenario for their healthcare: (1) They should consider their approach to technology adoption and whether they would like to encourage and facilitate for local, experience-based entrepreneurship; (2) They should consider the needs and wants of patients to access technology that can enable them. Based on this ideal future, policy-makers should consider laws and regulations and whether they can be altered for a more dynamic exchange between developer, user and the wider institutional arrangements. In addition, programmes could be developed to develop norms and knowledge that are more in concurrence with the ideal scenario.

Finally, we anticipate future scholars who are inspired by this case study will concentrate on the benefits and challenges of entrepreneurs who innovate in the public sector. There is a need for theoretical work built on more specific dimensions in a larger conceptual apparatus and empirical indicators that can be tested. Moreover, work needs to done on how to enhance a more ideal allocation of resources when it comes to public technology spending. Finally, there is a need to do larger-scale empirical studies to compare RI entrepreneurs in different institutional settings and look at how RI is constraining or enabling them in meeting the institutional wall.

REFERENCES

Aldrich, H. E. & Fiol, C. M. (1994). Fools rush in? The institutional context of industry creation. *Academy of Management Review*, 19, 645–670.

Alvesson, M. & Kärreman, D. (2007). Constructing mystery: empirical matters in theory development. *The Academy of Management Review*, 32(4), 1265–1281.

Alvesson, M. & Sandberg, J. (2013). Has management studies lost its way? Ideas for more imaginative and innovative research. *Journal of Management Studies*, 50(1), 128–152.

Barron, D. (1998). Pathways to legitimacy among consumer loan providers in New York City, 1914–1934. *Organization Studies*, 19(2), 207–233.

Bate, P. (2000). Changing the culture of a hospital: from hierarchy to networked community. *Public Administration*, 78, 485–512. doi: 10.1111/1467-9299.00215.

Berger, P. L. & Luckmann, T. (1966). *The Social Construction of Reality: A Treatise in the Sociology of Knowledge*. New York: Doubleday.

Bitektine, A. (2011). Toward a Theory of social judgments of organization: the case of legitimacy reputation and status. *The Academy of Management Review*, 36, 151–179.

Borins, S. (2001). The Challenge of Innovating in Government. https://www.researchgate.net/publication/242172041_The_Challenge_of_Innovating_in_Government (accessed 15 March 2019).

Boyd, R. (1979). Metaphor and theory change: what is "metaphor" a metaphor for?, in Ortony, A. (ed.), *Metaphor and Thought*. Cambridge: Cambridge University Press, pp. 356–408.

De Vries, H., Bekkers, V. & Tummers, L. (2015). Innovation in the public sector: a systematic review and future research agenda. *Public Administration*, 94(1), 146–166.

Delbridge, R., & Edwards, T. (2013). Inhabiting institutions: Critical realist refinements to understanding institutional complexity and change. Organization Studies, 34(7), 927–947. http://dx.doi.org/10.1177/0170840613483805

Feldman, M. S. (1995). *Strategies for Interpreting Qualitative Data*. Thousand Oaks: Sage Publications.

Giddens, A. (1984). *The Constitution of Society*. Cambridge: Polity.

Gioia, D. A., Corley, K. G. & Hamilton, A. L. (2012) Seeking qualitative rigor in inductive research: notes on the Gioia methodology, *Organizational Research Methods*, 16(1), 15–31.

Hannan, M. T. & Freeman, J. (1977). The population ecology of organizations. *American Journal of Sociology*, 929–964.

Heracleous, L., & Barrett, M. (2001). Organizational change as discourse: Communicative actions and deep

Holm, S., Liss, P. E. & Norheim, O. F. (1999). Access to health care in the Scandinavian countries: ethical aspects. *Health Care Analysis*, 7(4), 321–330.

structures in the context of information technology implementation. Academy of Management Journal, 44 (4),

697–713. http://dx.doi.org/10.2307/3069414

Hunt, S. D. & Menon, A. (1995). Metaphors and competitive advantage: evaluating the use of metaphors in theories of competitive strategy. *Journal of Business Research*, 33, 81–90.

Huy, Q. N. (2001). Time, temporal capability, and planned change. Academy of Management Review, 26(4), 601–623. http://dx.doi.org/10.2307/3560244

Kostova, T. & Zaheer, S. (1999). Organizational legitimacy under conditions of complexity: the case of the multinational enterprise. *The Academy of Management Review*, 24(1), 64–81.

Meyer, J. W. & Rowan, B. (1977). Institutionalized organizations: formal structure as myth and ceremony. *American Journal of Sociology*, 83(2), 340–363.

Mørland, B., Ringard, Å. & Røttingen, J. A. (2010). Supporting tough decisions in Norway: a healthcare system approach. *International Journal of Technology Assessment in Health Care*, 26(4), 398–404.

Oftedal, E., Iakovleva, T. and Foss, L. (2017). University Context Matter: A New Institutional Perspective on Entrepreneurial Intentions in Students. *Special Issue for Education and Training.*

North, D. (1990). Economic performance through time. *American Economic Review*, 84, 359–368.

Palthe, J. (2014). Regulative, normative, and cognitive elements of organizations: implications for managing change. *Management and Organizational Studies*, 1(2), 59.

Potts, J. & Kastelle, T. (2010). Public sector innovation research: What's next?. *Innovation*, 12(2), 122–137.

Rittel, H. W. J. & Webber, M. M. (1973). Dilemmas in a general theory of planning. *Policy Sciences*, 4(2), 155–169.

Scott, R. (2014). *Institutions and Organizations*. Thousand Oaks, CA: Sage Publications.

Scott, W. R. & Meyer, J. W. (1991). The rise of training-programs in firms and agencies: an institutional perspective. *Research in Organizational Behavior*, 13, 297–326.

Stilgoe, J., Owen, R. & Macnaghten, P. (2013). Developing a framework for responsible innovation. *Research Policy*, 42(9), 1568–1580.

Suchman, M. C. (1995). Managing legitimacy: strategic and institutional approaches. *Academy of Management Review*, 20, 571–610.

Vickers, I., Lyon, F., Sepulveda, L. & McMullin, C. (2017). Public service innovation and multiple institutional logics: the case of hybrid social enterprise providers of health and wellbeing. *Research Policy*, 46(10), 1755–1768.

Weber, A. P. & Khademian, A. (2008). Wicked problems, knowledge challenges, and collaborative capacity builders in network settings. *Public Administration Review*, 68(2), 334–349.

11. The role of user-led regional innovation networks in shaping responsible innovation in eHealth: lessons from the East of the Netherlands

Kornelia Konrad, Verena Schulze Greiving and Paul Benneworth

11.1 INTRODUCTION

Technology is increasingly penetrating every sphere of human activity, including health and social care. Many expectations are being placed upon eHealth – healthcare practices delivered or supported by information and communication technologies – to deal with a range of challenges faced in providing medical coverage to an aging population. The term eHealth has become an umbrella term for a wide spectrum of technologies that vary greatly in cost, autonomy and complexity. This ranges from simple online systems for managing doctor appointments and prescriptions, through technologies for sensing vital functions and monitoring a person's lifestyle and fitness level, to complex online communication and sensing platforms (Peeters et al., 2016; Kos, Sedlar and Pustisek, 2016). Technological interventions via eHealth are expected to deliver a range of goals, including increasing access to healthcare, improve quality, efficiency, safety and management of the delivered healthcare, reducing treatment costs, and augmenting patient self-management capacities for elderly and chronically ill patients (Dimitrova, 2013; Horn et al., 2016; Peeters et al., 2016).

However, policy-makers' enthusiastic promotion of eHealth has not matched its uptake and implementation in primary healthcare (Peeters et al., 2016). eHealth has been plagued by concerns regarding user safety, data security and privacy, alongside problems facing General Practitioners, whether practical (Internet connectivity), expertise to run these systems, technological and financial constraints (Peeters et al., 2016). Embedding and aligning eHealth innovations with existing healthcare systems that are highly regulated and vary considerably between places has proven costly and complex, involving many different actor groups. Embedding and aligning actors is critical to implementing eHealth, demanding detailed consideration of short/long-term consequences for involved actors' needs, expectations, and practices. In short, eHealth is a perfect domain to consider how innovations can be developed responsibly, including different stakeholders, anticipating and reflecting on potential implications, and using these insights in development and implementation. In particular, we are concerned with this volume's central question, namely who should be involved and in which form in the design, implementation and societal embedding of these emerging systems and practices.

In recent years, Responsible Research and Innovation (RRI) has emerged as a concept and governance approach aimed at aligning research and technological innovation with the needs and expectations of society (Owen et al. 2013; von Schomberg, 2013). Much attention has been paid to how researchers and innovation actors may become more aware of societal expectations, of different stakeholder groups, how they can anticipate possible impacts and integrate these insights in the research and innovation processes. However, accounting for the perspectives of different stakeholder groups in a top-down manner can be challenging; alternatively, responsible practices can be enacted at a grass-roots level to allow local communities to shape the technologies that affect their lives and for which their implicit consent is given. In this chapter we therefore ask the research question: *What possibilities exist for local communities to contribute to responsible innovation processes?* We do this by following the implementation of one specific eHealth application and exploring how far the underlying innovation processes have been responsible. We take a single case study of an eHealth system developed to coordinate communications between health professionals and patients, exploring (1) how different dimensions of responsibility were present during the innovation process, and (2) whether local bottom-up networks were conducive to enacting dimensions of responsibility.

11.2   CONCEPTUAL FRAMEWORK

Concepts of responsible research and innovation have arisen in part as a response to an increasing societal unease with the pace and consequences of technological change (Ribeiro et al., 2018). This unease can be linked to uncertainties about possible uses and impacts of emerging technologies, as well as the increasing complexity and interconnectivity of many (socio-) technical systems that exacerbate the difficulty to anticipate in advance the societal effects of particular interventions and facilitates second-order effects and uses beyond initial intentions. Above all is a sense that society has little control of these technologies allowing them to be implemented in undesirable ways. This is exemplified in the ongoing discussions regarding technology companies extracting value out of user data rather than purchased functionality, something clearly potentially problematic in health innovation.

11.2.1 The Rise of Responsible Research and Innovation as a Governance Process

These challenges are by no means novel, although the intensity and awareness appears to have latterly increased. In the 1960s, there was interest in the democratization of technology, particularly in north-western Europe, in developing tools and approaches that allowed communities to determine the conditions under which new technologies would be launched. In response to a number of crises of confidence around emerging food technologies, such as genetically modified food and growth hormones for dairy cattle, public understanding of science (PUS) emerged to better inform publics about the reality of the risks and opportunities of new technology developments. The 'deficit' model ('the public would agree with technology if only they understood it') implicit in PUS was rapidly critiqued, leading to more emphasis on co-creation and engagement with publics in research and innovation processes. Responsible research and innovation can therefore be seen as the latest in a lineage of concepts attempting to understand the ways in which societies value and consent to technologies. RRI seeks to provide tools to shape technologies in socially appropriate and desirable ways, and to eventually contribute to stronger, smarter and more socially just societies. As Stilgoe, Owen and Macnaghten define it (2013, p. 1570):

> Responsible innovation means taking care of the future through collective stewardship of science and innovation in the present.

In this chapter, we recognize that the concept of Responsible Research and Innovation is an emerging one, and not fully conceptually stable, whilst having a number of portmanteau characteristics not fully worked through. RRI's basis is that 'responsibility' in an innovation emerges when societal actors have opportunities to enact repertoires that influence innovators' and immediate beneficiaries' choices in translating an idea into a launched product, service or technique. Stilgoe et al. (2013) suggested anticipation, reflexivity, inclusion and responsiveness as key dimensions of responsible innovation. Anticipation refers to techniques and practices, including systematic procedures, envisaging possible impacts, relevant developments and opportunities, including a reflection on what is deemed plausible, more or less likely, and potential alternatives. Anticipation helps making choices mindful of future uncertainties and implications, allowing different stakeholders to express their preferences regarding potential trade-offs. Reflexivity involves considering one's own role, activities, value system, pre-assumptions and framings, and reflecting on the consequences that one's innovation process choices have for others, including those with other world views and value systems. Inclusion involves mobilizing forums where stakeholders and citizens come together and transform dialogue into meaningful influence choices made affecting their wellbeing. Responsiveness involves innovation products and processes actually being attuned to signals coming from different stakeholders, and emerging from considering the other three dimensions.

Stilgoe et al. (2013) propose a set of normative process characteristics that should supposedly facilitate research and innovation processes and eventually products that address different stakeholders' needs, values and concerns. They also exemplify various techniques and approaches applicable to research and innovation to foster anticipation, reflexivity, inclusion and responsiveness, such as foresight, codes of conduct, focus groups or value-sensitive design (Stilgoe et al. 2013, p. 1573). The 'indicative techniques and approaches' listed are all structured processes and procedures dedicated to achieving goals in line with RRI's suggested dimensions. In our study, we apply an open understanding to the types of processes and practices conducive to realizing RRI's dimensions, including systematic, dedicated processes alongside practices which may or may not be geared towards realizing specific dimensions, but in practice contribute to them.

Stilgoe et al. (2013) also question whether consideration is necessary for the institutional conditions and innovation system's characteristics enabling or constraining the suggested RRI dimensions. This concern for understanding the conditions for RRI also emerges in Walhout et al.'s (2016) approach. They suggest to study processes of 'RRI in the making', indicating there are many existing de facto practices, processes and governance arrangements in current research and innovation systems contributing to features of responsible innovation, irrespective of whether these explicitly refer to the concept. Furthermore, they demand further attention for how existing actor landscapes, governance arrangements and practices condition such RRI in the making (Walhout et al., 2016, p. 48). This perspective foregrounds the specific actor constellations, governance arrangements and practices within which innovation processes unfold. Thus, following Walhout et al. (2016), we contend more attention is required for different kinds of contexts where 'responsibility in the making' is evident.

11.2.2   The Regional Dimension to Responsible Research and Innovation

Innovations unfold within networks that embody their own internal governance processes; these networks are themselves embedded within a wider, more general landscape of regional, national and international (e.g., European-level) regulatory and administrative regimes. This is particularly apparent in the domain of healthcare where regulatory systems, organizations, practices and actor constellations differ clearly between countries, and sometimes sub-national territories. Furthermore, it has long been recognized that locality influences innovation processes (Alderman and Thwaites, 1992) and more latterly that an important role is played by the wider landscape of innovators, intermediaries, knowledge suppliers, policy-makers and skilled workers (Cooke, 2005).

The various kinds of proximity provided by co-location within a region facilitate innovating actors working together more easily with each other and complementary actors (Boschma, 2005). This ease of interaction facilitates repeated interactions, which may acquire network properties ('I can collaborate more easily with you because I know your partners') and systemic properties ('I can collaborate more easily with an unknown actor because they are located in the same region as me'). Contemporary regional innovation literature has treated innovation neutrally or positively, rationalized as innovation raising total factor productivity, thereby contributing to regional growth and improving quality of life (Temple, 1998). What has received far less attention is the governance arrangements in these networks that arise when policy-makers take decisions that shape innovators' capacities to innovate.

It is possible to regard these regional innovation networks as potential sites where various repertoires of responsibility may play out, and thereby including local and regional concerns in decision-making, and where proximity and close ties may facilitate some of these repertoires. The use of regional innovation platforms or coalitions deciding on regional priorities could potentially provide an anticipatory space where local users can reflect on what wider development trends might mean for the implementation of particular technologies in these local contexts. Reflexivity may correspond with the input of external expertise into regional innovation strategy development processes as recommended in standard strategy guides (e.g., Foray et al., 2012, 2009), while there may be a risk to take regional concerns and values for granted. Proximity and close ties may facilitate the inclusion of some types of stakeholders, and ease routes for feedback and eventually responsiveness. Still, efforts will still be necessary to ensure that any discursive processes involve people from a range of backgrounds and with a range of value systems. Responsiveness may furthermore emerge in the ways that policy-makers incentivize and reward innovators for pursuing innovations in ways that are reactive to local societal challenges, norms and value systems.

We argue that these regional innovation networks and communities could potentially provide a governance context where it is specifically possible to explore this issue raised by Walhout et al. of 'responsibility in the making'. We study the emergence and implementation of an eHealth platform, which emerged from a local network of healthcare actors in the Dutch province of Overijssel, and subsequently diffused along regional networks. The platform was created in response to an exogenous national health system change to ensure continuity and consistency of service provision for elderly vulnerable residents. Our study asks two operational research questions:

1. To what extent this innovation process exhibited characteristics of responsible innovation, and
2. Whether/in which form local/regional communities and local/regional context were influential.

We refer to local communities, networks, and contexts if these extend predominantly in a municipality. We refer to regional networks and contexts, if they extend beyond a single municipality, but still build on immediate geographical proximity. We first present the innovation process of creating the eHealth portal, to subsequently examine how far regional innovators were able to enact repertoires of responsibility within the overall innovation process, and to which extent local and regional network and governance aspects were influential.

11.3 METHODOLOGY AND INTRODUCTION TO THE CASE STUDY

11.3.1 Study Methodology

In this chapter we are concerned with whether local and regional networks influence the exercise of responsibility repertoires in an eHealth innovation process. To answer this question, it is therefore necessary to gain relatively in-depth, detailed knowledge about one or more cases that provide sufficient information to make valid claims about the ways that responsibility repertoires have been exercised, and to relate that back to networks, interests and roles of local actors as well as specific characteristics of those places and settings. We have therefore chosen a qualitative case study approach, to produce sufficient depth of understanding about that situation.

The case study, which we here refer pseudonymously to as CareConnect, builds on 10 semi-structured interviews which were carried out with key stakeholders in the development of the overall system in the summer of 2017. Seven of those directly related to CareConnect: a key innovator, a software developer, representatives of a municipality, an insurance company, and a nursing service, a general practitioner, and the project leader of a collaborative communication network in the mental health care sector. Additionally, three actors with expertise on the regional eHealth innovation system were interviewed. Interviews lasted about one to one and a half hours, were recorded and paraphrased. We analysed the data to explore how the innovation process unfolded, starting from the initial idea's emergence, the first experimental implementation, through to the development into a full functioning system being upscaled to other localities.

11.3.2   The Actor Landscape: CareConnect in the Dutch Healthcare System

Organizationally, CareConnect was established in 2013 as an independent foundation. At the time of the research, around 400 organizations and institutions were actively involved in CareConnect with the platform operational in more than half of Overijssel Province's municipalities. CareConnect retains operational autonomy through financing via service reimbursement by municipalities and insurance companies. CareConnect is an online communication platform that creates a network around the client (patient), linking caregivers, family members and caretakers. This platform supports elderly people (and those with chronic diseases requiring regular help/care) to live longer at home, by enabling better communication and better aligning different parties involved in a particular patient's care. CareConnect is embedded in a website where the patient or his/her caretaker can ask a question, which is directly send to other parties in the network who can answer this question in a short time. Parties in the network include pharmacies, homecare, physiotherapists, hospitals, general practitioners and the municipality (see Figure 11.1). Typical messages sent via CareConnect are updates on the used medication, orders at the pharmacy or questions about care. In addition, a care plan can be defined in the system. This way, the involved parties are updated about the care of a particular person and about the actions of other actors with the aim to increase the quality and efficiency of the care.

In the Netherlands, health insurance is accessible and obligatory for everyone. Since 2006, the Dutch health insurance consists of a basic package that covers general needs and which is determined by the ministry of public health, welfare and sport (VWS) after consulting the Dutch care institute (ZiN) (see Figure 11.2). Local policy-makers, health professionals or intended users thus have little influence on what is covered by this basic health insurance. For specific services, the insurance company can buy additional packages. Services

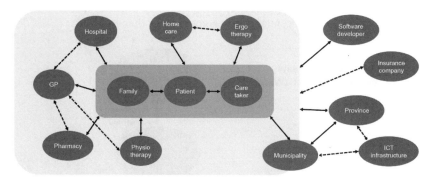

Figure 11.1   The network of actors related to CareConnect

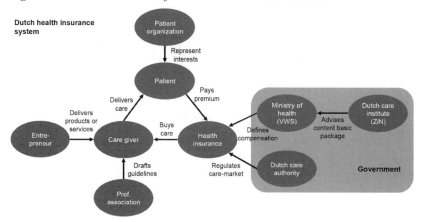

Figure 11.2   The Dutch health insurance system linking patients to care providers

are offered by caregivers who can make special deals with insurance companies. These deals lead to a high competition in the market between caregivers and insurance companies and offers some choice for insured persons. Next to this health insurance law, there is a specific law for societal support. This law regulates the support for people who are living at home but are dependent on regular care (e.g., help in the household, more protected living, mobility aids, re-integration, and so on). For this particular law, the municipality, instead of the government, is responsible for the funding of care and takes over the role of the health insurer, selects services and makes contacts with caregivers as shown in Figure 11.3 (Janssen, 2014). In this way, local actors have an important role in defining how care services are provided. Furthermore, Dutch health insurers are by custom and practice more regionally organized, with specific insurers playing a more central role in some regions than in others.

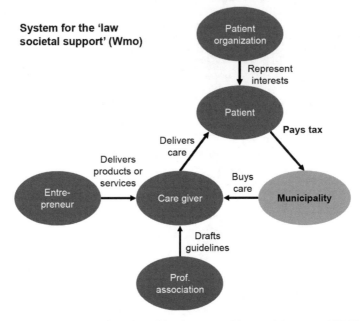

Figure 11.3   System introduced by the Law of Societal Support (WMO)

11.4   THE CARECONNECT INNOVATION PROCESS

In this section, we present an overview of the innovation journey through which CareConnect has unfolded. First, the emergence of an idea for a platform along with a consensus for action is described. Second, we describe the guiding design principles and how user feedback was incorporated. Third, we address the launch of the platform amongst the initial user group, followed by the subsequent operation and expansion amongst a number of wider users. This is not a fully sequential process, as development and adaptation continued throughout implementation and expansion. This section provides the basis for a subsequent analysis in which we look at the extent to which it was possible to observe responsible repertoires across these four stages. This then provides the basis upon which we can answer our operational question and contribute to the wider discussion within this volume as a whole.

11.4.1   The Emergence of the CareConnect Concept

The idea of CareConnect originated in a small village in the East of the Netherlands in response to a 2011/2012 change to Dutch healthcare regulations which raised the eligibility criteria for subsidized care home places for elderly residents. This reduced the demand for care home places across the Netherlands, and led to the closure of the care home in this village at the end of 2012. In the village, that care home had worked closely with a local health centre, which included general practitioners, nursing services, a physiotherapist and a pharmacy, who were able to coordinate with each other when providing more complex care services to care home residents. In late 2012, when the care home was to be closed, the carers discussed among themselves and with others what exactly was considered so difficult about having elderly clients staying longer at home. They concluded that the key problem was a lack of communication and coordination of the activities of the different parties involved in the care. In addition, the newly appointed carers at home, often relatives (the so-called 'mantelzorgers'), were typically not integrated well in the communications around the client and didn't know well whom to address in case of need. Actually, the family carers were usually more concerned and in need of communication than the clients themselves. Using email as a means of communication was considered too insecure and reaching others by phone too time-consuming to guarantee that patient-critical knowledge would be effectively shared. Thus, as a potential solution a digital communications platform was envisaged.

From the beginning, the initiators of the platform determined that the patients should be central to the system and own all their data. This decision was not the result of a long discussion or a systematic consultation with other stakeholders, but an approach that felt intuitively right to take, not only for reasons of ensuring the autonomy of the patients, but also due to functional-pragmatic considerations and to avoid taking sides with one particular party, potentially discouraging others from joining. Existing systems, for instance of GPs or physiotherapists, are directed at the needs and interests of the particular professional group, but are not oriented at facilitating the communication in a heterogeneous care network. Hence, they may include data that are not appropriate for sharing, or the information lacks elements necessary from the perspective of the patient or care network. However, many actors involved in discussions, such as further GPs, were also sceptical, fearing that the use of the system would be too time-consuming.

An independent foundation was established, CareConnect, which ensured that it remained organizationally separate from the other care providers who all had their own care systems. The establishment of the foundation was followed by creating a supervisory board with municipality representatives, regional nursing services in the region, carers and GPs to ensure the system fitted into the overall Dutch health system. Furthermore, team members and advisors of the foundation mostly have a background in various roles in the care sector. The foundation received important support in the organizational and administrative set-up from a local resident providing his expertise on a voluntary basis. The foundation initially received a small subsidy from a regional care service organization, the local municipality and a local nursing service, to run a pilot in a part of the municipality. The subsidy also provided CareConnect with the aim to extend it to the whole municipality if the pilot turned out successful. During an official ceremony, CareConnect was opened by the municipality Director, with the first client present.

11.4.2   The Adaptation of Existing Software

Once the idea for a digital platform had emerged, the consortium sought an off-the-shelf solution that could meet their various requirements. The foundation searched on the Internet for potential tools, and identified a platform that had already been developed by a software developer who had himself provided care for his disabled daughter and had learned about communication struggles between carers and care professionals experienced by friends. His platform provided a solution to this problem by allowing messaging services between the client, family and care providers, and by providing the possibility to set up a care plan for the client as a more systematic way to communicate and coordinate the necessary care actions.

Initially, CareConnect used the software in its original form, but it soon became apparent that clients and the nursing service used the system differently. Examples of these deviations included clients uploading daily blood sugar curves in the system, or nursing services placing pharmacy orders via the system. Following the actual use of the system and the feedback of the users, the system was adapted to facilitate those interactions.

A key decision forum in the evolution of the platform related to the choice of which functionalities would be added to the system. CareConnect advocates strictly a 'less-is-more' approach, allowing a new client record to be opened simply by linking a patient, carer and GP in the system. This record can later be augmented by creating a care plan or by adding additional users, allowing the system to be set in place as quickly as possible. This approach differed clearly from those of former clients of the software developer, who first required to enter a complete care plan, which, however, often prolonged the process of getting a new client account operational. Simplicity in the system design, however, was not easy to achieve, as the system has to be accessible for very diverse user groups. Elderly clients and their carers often have a low digital literacy and little knowledge of the medical terms, whilst young health professionals by contrast tended to have a high technological affinity and a professional medical education. This implied that the software development required much more effort in designing for simplicity than for additional functionalities.

In the development process, the project director of the foundation served as an intermediary between the software developer and the users. When potential improvements were proposed on the basis of user feedback, they were incorporated into a concept version that was then piloted and tested in a number of locations. If the modifications were positively evaluated after the pilot, they were translated into a software update that was released to all users. Furthermore, problems were regularly identified at the annual user meeting, bringing together all people that coordinate the implementation of the CareConnect system (e.g., district nurses, GPs, or employees of CareConnect). These meetings were arranged to encourage more people to provide feedback to the Director–Developer group, as well as to allow users to exchange experiences and facilitate peer learning.

Prior to CareConnect becoming a client of the software, the software had been acquired by a company that was already in the market of healthcare software and systems, and thus better equipped than the developer to handle the juridical and commercial aspects. In the meantime, all sales and first line customer support is handled by CareConnect, as they are better able to facilitate the local cooperation and communications necessary between the various parties involved when setting up a new application of CareConnect than the developer or the company owning the software. The software developer is mainly contacted for specific technical problems and development, and the healthcare systems company advises on the growth strategy and provides support through their network of contacts with GPs and pharmacies.

11.4.3   Implementation and Expansion

The implementation of CareConnect started in 2013/2014 with a pilot in the above-mentioned village, and it was then broadened to the municipality the village is part of. These first steps were supported by a small subsidy from a local care organization, the municipality and a nursing service. Since 2015 CareConnect has subsequently spread to more than half of the municipalities in the province, and to a smaller degree beyond. Initially this was supported by a public research grant, investigating the conditions for upscaling, and the benefits and costs for the different user groups, with the further effect of raising broader attention for CareConnect. Already the first expansion step from the village to the municipality built on the support of local networks. Initially there was scepticism amongst GPs in the municipality, but the system was championed by the local out-of-hours practice which had encountered CareConnect in the pilot phase. These doctors were sufficiently positive about their experiences and the benefits for care as to persuade other providers to be willing to do it. The kinds of positive experiences that were related by this out-of-hours group related to the fact that it was regarded as saving both time on an ongoing basis as well as demanding less time to be implemented. The vector for the spread of these messages was not formal, but rather came through the different employments that a number of these doctors had, working in other practices or policlinics alongside their out-of-hours work. Certainly, there was little active promotion by the foundation of the platform, although they did find themselves responding to requests for more information from people in other regions who had heard (positively) of the system and sought to introduce it in their municipality.

One of the key decisions taken with the roll-out and expansion of the product to other municipalities outside its 'home location' is the agreement of a reimbursement compact. As explained above, according to the Dutch health-care system the reimbursement of health services for elderly in need of regular care is partly determined by the regionally organized health insurances and partly by the local municipalities. A health insurer involved early on, decided in 2015 to allow for reimbursement of costs for the system via the GPs, and considers further possibilities involving the care personnel, whereas munici-palities support the CareConnect foundation directly. As a key condition for starting the implementation of the system, CareConnect, the municipalities and health insurers required that all relevant parties in a local care network would be willing to communicate with each other and join the platform. Similar to the initial experience, it was the GPs who often were decisive in this process, because when GPs either actively promoted or were passively willing to use the platform then it proved easier to get other parties involved. Respondents noted that this related to a particular feature of the region, with its strongly rural character, which was that the GPs often had a dominant position in the care network and often accompanied a patient and whole families throughout their lives. Alongside GPs, nursing services were also important drivers for the implementation, and in those villages where the nursing service was very busy or short staffed then implementation typically progressed very slowly.

11.4.4 CareConnect Operating in Practice

We will now consider the extent to which the system has been able to live up to the initial intentions and promise to create a client-centred care system that leads to better patient treatment by facilitating better communication between multiple care providers involved in providing care for a particular patient. In this subsection, we consider the ways that the interests of three groups have become implemented through the innovation process, namely those respon-sible for organizing and financing the provision of care (municipalities and health insurers), healthcare professionals, and the recipients of care (clients and their families).

Municipalities and health insurances regarded the system as very positive. Health insurers have been extremely worried about the implications of a growing elderly population (along with limited opportunities for premium growth) since their creation in 2006, recognizing the time intensiveness of elderly care and thus the additional costs this will bring for GP-led care and the need for cooperation. However, the care sector has been relatively poor at proposing innovative solutions, and one of the features of CareConnect that attracted the interest and attention of one health insurer was precisely this feature, as an innovation emerging from healthcare practitioners and addressing this problem. The strong position of both clients and family members, was also considered a very strong point, by representing the legitimate interests of the clients, thereby facilitating the acceptability of the system.

In a similar way, one of the municipalities that joined CareConnect later on was specifically attracted by the fact that the initiative was supported bottom-up by the GPs and promised to facilitate the communication with this group that was considered as highly important for the care process. In addition, family carers were very positive about the system, a group of carers' municipalities had become responsible for after a change of law in 2007. The municipality appreciated in particular the effect it had on ensuring care quality and reducing the need for emergency respite care or cover to fill or repair short-term health problems arising from communication problems. According to the head of the responsible department, some employees of the municipality complained about not being included in the CareConnect network by some of the clients, whilst she considered this choice as a legitimate right of the patients and saw the task with the municipality to convince clients of the usefulness of adding the municipality to the network where necessary.

In terms of the *healthcare professionals*, the system embodied a number of their wishes. GPs found that it helped to alleviate work pressure, as they could complete reporting off-site and at times which were suitable to them, it saved time in communicating with pharmacies, and it allowed to substitute some face-to-face and phone encounters which could be handled more efficiently via CareConnect. The system is also perceived as improving the quality of care, as it facilitates consultation with a specialist before finalizing a report and because it allows a more intense communication with the family carers, which was only scarce before. It was, however, also mentioned that setting up the system and convincing all parties to participate is a time-intensive task. Finally, the interviewed GP experienced strong differences in the interest to engage with ICT-based innovations between the different regionally organized care organizations.

Likewise, the nurses found the system useful. The nurses certainly appreciated the fact that they were better informed prior to arriving at patients, that it facilitated handling situations which required action within short time, that it facilitates the communication exchange between different people involved in the care network who do not meet regularly. In particular, the contact with family members typically arriving when the professional carers are gone, has improved. Before, communication happened mainly in moments of crisis, whereas now family carers can share their observations and concerns. The system furthermore helps to avoid the loss of information, which may happen if only transmitted via different parties, and to receive information from pharmacies and doctors when needed.

The *care recipients'* experience of CareConnect was somewhat patchier, because they tended to consult the system less regularly, for instance at moments of change or uncertainty. However, the system did bring changes that were appreciated by the clients, such as allowing them to follow the communications between the care providers (who were the more active system users). The system also provided a mechanism to allow clients to empower their next-of-kin to also track care provision; previously, the client's family were dependent on a client's memory and comprehension of what they had been told to relate their treatment back to their family whilst the CareConnect system allowed clients to empower their family to view communications and their care plan. An additional advantage that emerged was that it allowed for small issues to be communicated between nurses, clients and families; these were important in situations such as identifying at an early stage behavioural irregularities around medicine intake or physical activity that might otherwise be unnoticed and which if necessary can then be translated into a notification for the attention of the GP. As a limiting factor, some of the clients do not use computers, but can still be involved in a more passive mode when carers show the messages they send or the communications which have occurred. There is, however, also a group of potential clients that is not interested in using the system.

11.4.5   Future Perspectives

There are a number of possible developments of the platform which are explored or discussed as possible future directions. Currently, new types of settings beyond elderly care are being explored and tested, such as psychiatric care and youth care, each posing different challenges, requirements and questions, as the typical structure of the care networks and the types of cares differs. In the case of youth care, it is furthermore not evident if the system should be organized around the child or the family. In addition, making the system more accessible for particular user groups, such as low literate people, adding new types of parties such as hospitals, and integrating additional features such as video calling or the integration of measuring devices are explored. It has been stressed in the interviews that an extension of functionalities may not so much be a technical challenge, but that it requires careful considerations on the use value for CareConnect, with a view on the current and future situations. As a general vision, most interviewees envisage the availability of an independent communication platform for the Netherlands, which may, but need not, be based on the CareConnect platform. In order to enable this, standards for data exchange, involvement of larger parties, and a clear system for reimbursements are considered as conducive.

11.5   DISCUSSION: ANTICIPATION, INCLUSION, REFLEXIVITY AND RESPONSIVENESS IN THE INNOVATION PROCESS OF CARECONNECT

In this section, we discuss if and in which form characteristics of responsible innovation are visible in the way the innovation process of CareConnect unfolded and in the way the system has been designed and used. In addition, we trace if and how the local/regional nature of the innovation network and the regional characteristics of the institutional environment have been influential.

At first sight, the strength of the innovation process of CareConnect is characterized by the ability to respond to and make use of imminent challenges and opportunities, rather than by explicit considerations about potential mid- to long-term future developments and impacts. An element of short-term anticipation is, however, visible from the very beginning. At the time the local impact of the national policy changes became apparent, the initiators of CareConnect did not wait to cope with the new situation until it occurred, but envisaged what would be the challenges and what might be ways to cope with them. Along the innovation journey, more systematic forms of anticipation have been triggered, still focusing on imminent challenges ahead, as the study conducted after the launch in the first municipality, investigating the conditions for upscaling the system, and the costs and benefits the platform was likely to entail for different user groups. At the time of writing, multiple directions for the further development are envisaged, also these perspectives emanated along the journey rather than being the result of dedicated anticipatory procedures. Actually, one of the interviewees advocated a more systematic approach towards reflecting on future possibilities. Thus, we conclude that overall in the CareConnect innovation process, anticipation played a moderate role; at the same time, we see that the anticipatory elements which did occur, clearly informed the development process, in this way constituting an element of responsiveness.

Reflexivity is arguably the most ambiguous of the dimensions, entailing multiple meanings (Stilgoe et al. 2013, p. 1571). Reflexivity in the form of awareness and reflection on the values guiding activities is clearly visible in the reflection on the core values guiding the design and organizational embedding of the platform, such as autonomy of the clients and patients respective user-centeredness. Interviewees stressed that also this consideration emanated rather intuitively from the initializing discussions among the core group of initiators, and not from a structured deliberative process. It should be noted, that the considerations for deciding to make these values central and the way these were translated in concrete technical and organizational decisions, built on a deep and varied knowledge of the use field incorporated by the core actors of CareConnect. Related to these core guiding values, we clearly observe mindfulness for the implications the platform has for different parties. Throughout the development and implementation process, dedicated attempts at learning about the perspectives of the different user groups have been undertaken. Furthermore, we find multiple instances where actors consider critically their own role and priorities, for instance when developers embrace the less-is-more approach in the design, or when municipality or health insurance employees embrace the principle of autonomy of clients, even if this may result in a certain loss of control for them. Overall, throughout the interviews we see that actors reflect on the effects the system has on own and others' practices. We argue that the close-knit, local and regional networks which carry the development and use of CareConnect are conducive to these forms of reflexivity, in particular for the more intuitive and interaction-based elements, as it facilitates becoming aware of the multiplicity of perspectives, values, roles and conditions of and impacts for different actors.

The RRI dimensions which are most evident in our case are inclusion and responsiveness. CareConnect can be considered as a case of a user-led innovation (Truffer, 2003; von Hippel, 2005). Actors who represent multiple – but not all – future user groups are the initiators of the platform, responding to what is perceived as a pressing societal and professional problem. Some of the initiators took over a key role in the foundation which carries CareConnect, and further actors rooted in the care sector joined the foundation in different roles. This bottom-up process developed further along networks characterized by proximity, co-location and organizational ties, in this way mobilizing a broader set of actors. Dedicated procedures for generating user feedback, particularly from those groups who are less likely to provide feedback proactively, and to allow for social learning among users are organized as well. While our methodological approach based on a small number of interviews can only provide limited or indirect evidence of the broader appreciation of CareConnect among the multitude of involved actors, it appears that precisely this inclusive and bottom-up approach is appreciated by many of the involved stakeholders and considered as an essential success factor, even by actors for whom this may imply a certain loss of control or increased complexity, as the already mentioned representatives of the municipality and health insurance. The expansion along existing social networks is also considered as important for creating interest among potential users that would else be sceptical about the use value of the platform.

Responsiveness is apparent as well, with the design having been adjusted continuously in line with the feedback received for use experience and feedback. It has been stressed by some of the interviewees that the platform is considered as rather flexible in allowing for adjustments. It has furthermore been highlighted that another important element in enabling the uptake and use of the system has been the fact that support for the system has been provided by the CareConnect foundation with its experience in the world of care, and close links to the different user groups, rather than by the owners of the software. The stepwise, bottom-up process, proceeding along comparatively small steps, with a focus on creating immediate use value rather than complex functionalities has furthermore facilitated quick implementation and learning processes.

11.6   CONCLUSION

To summarize, we see all four RRI dimensions put forward by Stilgoe et al. (2013) embodied in the innovation journey of CareConnect, though arguably to different degrees. It became also apparent that the regional, and partly local, structure of the actor networks carrying the innovation, implementation and use of CareConnect played a decisive role, both for the success of the system in a more general sense and the way elements of anticipation, reflexivity, inclusion and responsiveness featured in the process. At first sight, it may seem almost evident that a user-initiated innovation process shows a certain inclusiveness; still, we would like to highlight that inclusiveness was not confined to those participating in the innovation network, but that active steps were taken to involve relevant stakeholders more broadly. On the other hand, a stronger anticipatory approach, which for instance envisages in a structured way how further developments in the healthcare system could change the conditions for CareConnect to develop and discuss perspectives for new forms of usage, could possibly add relevant perspectives. While somewhat speculative, it may well be that precisely the focus on imminent needs and local conditions may have been conducive to taking such a perspective.

The regional and partly local character of most of the actors and networks involved in our case has been largely a result of the strong regional elements in the organization of the Dutch healthcare system, such as the responsibility of local municipalities for home care, the regional distribution of health insurers and the local and regional organization of a number of health services. It would be an interesting point for comparison, which is, however, beyond the scope of this chapter, to study to which extent and in which form regional innovation networks and processes unfold in differently structured healthcare systems.

Overall, we can conclude that the strong role of local bottom-up networks firmly rooted in the world of care facilitated the design and implementation of this eHealth application by targeting the application to user needs, mobilizing networks in the use domain throughout the implementation process, by allowing for flexible, rather agile experimentation and user involvement. In principle, one might expect that this approach should also enable adjustment to local variety, even though this has been less apparent in the process so far. In our case, the need for variety and tailoring the system to different conditions has predominantly been perceived with respect to applying the system to different care situations beyond elderly care.

We would furthermore like to highlight that the innovative potential of CareConnect is not so much linked to technical innovation, but rather to the societal challenge or problem, which has been identified as primarily an issue of coordination and information exchange, and the active societal embedding of the platform in use networks, practices and institutional frameworks. This is not to say that the specific technical design has not been an essential and complex issue, but complexity resided more in finding an appropriate design in an iterative process, rather than building on an innovative technology per se. This seems important to highlight against the background that despite the turn to societal challenges as a major reference point in research and innovation policy, it is still common, also within the context of RRI, to approach an innovation process with a focus on a particular technology or a focus on innovative technology (Kuhlmann and Rip, 2014). It seems remarkable, that when asked about future perspectives of CareConnect, a number of the interviewees pointed to the vision of a communication platform available on a national level, irrespective of the underlying technical system.

Finally, we would like to point out that the potential of improving healthcare by means of eHealth resides in this case in the societal organization of networks of patients and carers, rather than in monitoring and 'improving' individual patients or specific bodily functions of patients, a perspective which seems to be more common in many of the currently proliferating eHealth applications. We suggest that these social and organizational aspects of healthcare deserve broader attention in the development of eHealth.

REFERENCES

Alderman, N. & Thwaites, A. T. (1992). the regional dimension to the adoption of innovations. In Townroe, P. and Martin, R. (eds), *Regional*

Development in the 1990s: The UK in Transition (pp. 188–192). London: Jessica Kingsley.

Boschma, R. (2005). Proximity and innovation: a critical assessment. *Regional Studies*, 39, 61–74.

Cooke, P. (2005). Regionally asymmetric knowledge capabilities and open innovation: exploring 'Globalisation 2' – a new model of industry organisation. *Regionalization of Innovation Policy*, 34, 1128–1149.

Dimitrova, R. (2013). Growth in the intersection of eHealth and active and healthy ageing. *Technology and Health Care*, 21, 169–172.

Foray, D., David, P. A., & Hall, B. (2009). Smart specialisation: the concept. In *Knowledge for Growth: Prospects for Science, Technology and Innovation*. Selected papers from Research Commissioner Janez Potočnik's Expert Group, pp. 25–29. Retrieved 8 March 2019 from http://ec.europa.eu/research/era/pdf/knowledge_for_growth.pdf.

Foray, D., Goddard, J., Goenaga Beldarrain, X., Landabaso, M., McCann, P., Morgan, K., Nauwelaers, C. & Ortega-Argilés, R. (2012). Guide to research and innovation strategies for smart specialisation strategies. Sevilla: JRC. Retrieved 1 March 2019 from http://ec.europa.eu/regional_policy/sources/docgener/presenta/smart_specialisation/smart_ris3_2012.pdf.

Horn, G., Eliassen, F., Taherkordi, A., Venticinque, S., Di Martino, B., Bücher, M. & Wood, L. (2016). An architecture for using commodity devices and smart phones in health systems. *2016 IEEE Symposium on Computers and Communication (ISCC)* (June, pp. 255–260). IEEE.

Janssen R. (2014). *Innovatieroutes in de zorg. netwerkbijeenkomst 'decentraliseren=innoveren'*. Utrecht: Zorg voor innoveren.

Kos, A., Sedlar, U. & Pustisek, M. (2016). Research and innovation in ICT with examples in the field of eHealth and wellbeing. In Loshkovska, S. and Koceski, S. (eds), *ICT Innovations 2015: Emerging Technologies for Better Living* (pp. 1–10). Berlin: Springer-Verlag Berlin.

Kuhlmann, S. & Rip, A. (2014). The challenge of addressing Grand Challenges. A think piece on how innovation can be driven towards the 'Grand Challenges' as defined under the prospective European Union Framework Programme Horizon 2020. Report to the European Research and Innovation Area Board (ERIAB).

Peeters, J. M., Krijgsman, J. W., Brabers, A. E., De Jong, J. D. & Friele, R. D. (2016). Use and uptake of eHealth in general practice: a cross-sectional survey and focus group study among health care users and general practitioners. *JMIR Medical Informatics*, 4(2), e11.

Owen, R., Bessant, J., Macnaghten, P., Gorman, M., Fisher, E. & Guston, D. (2013). A framework for responsible innovation. In Owen, R., Bessant, J. and Heintz, M. (eds), *Responsible Innovation: Managing the Responsible*

Emergence of Science and Innovation in Society (pp. 27–50). Chichester: John Wiley & Sons.

Ribeiro, B., Bengtsson, L., Benneworth, P., Bührer, S., Castro-Martínez, E., Hansen, M., Jarmai, K., Lindner, R., Olmos-Peñuela, J., Ott, C. & Shapira, P. (2018). Introducing the dilemma of societal alignment for inclusive and responsible research and innovation. *Journal of Responsible Innovation*, 5, 316–331.

Stilgoe, J., Owen, R. & Macnaghten, P. (2013). Developing a framework for responsible innovation. *Research Policy*, 42, 1568–1580.

Temple, J. (1998). The new growth evidence. *Journal of Economic Literature*, 37, 112–156.

Truffer, B. (2003). User-led innovation processes: the development of professional car sharing by environmentally concerned citizens. *Innovation: The European Journal of Social Science Research*, 16, 139–154.

von Hippel, E. (2005). *Democratizing Innovation*. Cambridge, MA: MIT Press.

von Schomberg, R. (2013). A vision of responsible research and innovation. In Owen, R., Heintz, M. and Bessant, J. (eds), *Responsible Innovation* (pp. 51–74). Chichester: John Wiley & Sons.

Walhout, B., Kuhlmann, S., Ordonez Matamoros, H. G., Edler, J., Lindner, R., Randles, S. ... & Mejlgaard, N. (2016). ResAGoraA concepts and approach. In Lindner, R., Kuhlmann, S., Randles, S. et al. (eds), *Navigating towards Shared Responsibility* (pp. 47–53). Karlsruhe: Fraunhofer Institute for Systems and Innovation Research ISI.

12. Management of stakeholders' knowledge for responsible research and innovation

Elisa Thomas and Luciana Maines da Silva

12.1 INTRODUCTION

Reaching and involving stakeholders in knowledge sharing may be managed physically and in virtual environments (Yenicioglu and Suerdem, 2015). However, the coordination of participation-based roles is complex and time consuming, mainly when it comes to physical interactions. Given the importance of tacit knowledge for valuable inputs for R&D, current virtual platforms do not fully facilitate the exploration of stakeholders' participation by firms performing R&D.

Based on the premise of the need for inclusion of stakeholders in RRI (Stilgoe, Owen and Macnaghten, 2013) and considering that valuable knowledge for R&D may be shared in different ways, this study aims to analyse: How do firms manage stakeholders' knowledge through participation on a RRI context? To answer this question, our chapter focuses on the processes of innovation rather than on the outcomes (Burget, Bardone and Pedaste, 2017) and analyses it in two Brazilian start-ups. Social inequality increases in the poverty rate and difficulties in accessing basic public services such as health and education are characteristics of emerging markets such as Brazil. Even though, despite of the market's complexity, entrepreneurs see business opportunities in minimizing the difficult access to healthcare and in facilitating a more open and inclusive communication.

After this introduction, the chapter is organized as follows: we first draw on previous bodies of literature about RRI and knowledge management to establish the theoretical foundations of our work. Afterwards, we describe the method design and the context of the empirical research. Then we present two case studies and our main findings regarding the integration of tacit and explicit knowledge coming from stakeholders' participation in the RRI process. The empowerment of citizens takes place through their social inclusion, as provided by Hand Talk, and their power of choice, as provided by SOSPS. We finish the chapter with implications for practitioners and for the literature.

12.2   LITERATURE REVIEW

12.2.1   Responsible Innovation and Its Dimensions

Responsible Research and Innovation (RRI) is a process that includes stakeholders, anticipating potential implications, reflecting and responding to the needs and values of society. Overall, it comprises ethical considerations as well as widespread governance issues (Burget, Bardone and Pedaste, 2017). One of the main assumptions of RRI is the participation of multiple stakeholders, with the objective of better aligning the outcomes of R&D with the necessities and expectations of people (Lettice et al., 2017).

Chapter 2 of this book discussed the four dimensions which make a guideline for RRI: anticipation, inclusion, responsiveness, and reflexivity. This chapter focuses on **inclusion** which refers to dialogue, engagement, and debate among different stakeholders, a process that is centred around a quest for social legitimacy for innovation (Demers-Payette, Lehoux and Daudelin, 2016). Also, we look at **responsiveness**, which discusses the way firms in RRI contexts need the ability to adapt the form or direction of its innovation trajectories in response to stakeholders' values and reactions.

12.2.2   Types of Stakeholders' Knowledge: Tacit and Explicit

The knowledge-based view of the firm together with open innovation literature stresses the value of external knowledge for innovation added to the internal knowledge of the firm (De Silva, Howells and Meyer, 2017). Also, in RRI processes, external knowledge can be acquired in tacit and explicit forms.

For a couple of decades, knowledge for innovation has been studied by scholars and managed by firms considering two different complementary characteristics: (1) tacit knowledge as the collection of experiences, intuitions and skills (Leonard and Sensiper, 1998; Nonaka and Takeuchi, 1995; von Krogh, Ichijo and Nonaka, 2000); and (2) explicit knowledge as information easily communicated in formal language and made available to people in the firm (Nonaka, von Krogh and Voelpel, 2006; Smith, 2001). Factors such as reciprocity, enjoyment, and social capital contribute significantly to enhancing employees' tacit and explicit knowledge sharing intentions (Hau et al., 2013).

More recently, the integration of these two types of knowledge has shown to happen not solely indoors, but also with external partners on innovative processes, despite the risks involved in knowledge sharing across company borders. In alliance partners, Becerra, Lunnan and Huemer (2008) found that sharing explicit knowledge is closely associated with the firm's willingness to take risk, but sharing tacit knowledge is strongly related to high trustworthiness. Despite existing criticism towards categorizing knowledge as belonging to either one or the other class (Jasimuddin, Klein and Connell, 2005), we take the approach of tacit and explicit knowledge having different characteristics for the purpose of better identifying the way they are managed in firms.

12.3 METHOD

The research follows a qualitative research strategy through case studies. As RRI is a fairly new concept for firms, and the dimensions of inclusion and responsiveness are raised as vital concerns for RRI processes, qualitative data helped us to understand how these concepts are integrated with knowledge management in practice. The case studies were purposefully selected seeking for information richness.

12.3.1 Context of Empirical Research

We set our research in Brazil, where the context of an emerging country could provide insights about a different setting from most research on RRI. Brazil is a country of continental dimensions, with 8.5 million square kilometres, with more than 208 million inhabitants (IBGE, 2018). More than 80 per cent of the population inhabits less than 1 per cent of the national territory, resulting in regions with high levels of population density. The city of São Paulo has more than 13,000 inhabitants.

The Brazilian Constitution determines that health is the right of everyone and the duty of the State. To meet this determination, the Unified Health System (Sistema Único de Saúde – SUS) was created in 1988 (Constituição do Brasil, 2001). SUS is financed with funds from the social security, the Federal, State, Federal District and Municipalities budged. Although it is not able to provide a quality service, due to the federative model (De Souza, 2002), corruption, as well as scarcity of resources. These problems result in deficient and time-consuming care, causing 22 per cent of the population to seek private healthcare (ANS, 2018). However, even using private plans, the hospital system is shared with SUS, which generates overcrowding, delays and high costs. The latest survey dating back to 2009 shows 2.4 hospital beds per thousand inhabitants in Brazil, while the World Health Organization (WHO) recommends from three to five beds per thousand inhabitants (ANAHP, 2018). Besides, patients can wait for more than two years for a medical appointment through SUS (TelessaúdeRS, 2018).

Developed countries may face similar problems. However, emerging economies have several risks, such as weak political institutions, corruption, social tensions, overuse of natural resources and environmental degradation (Ramamurti, 2012) that imply the need for RRI. On the one hand, emerging countries such as Brazil present many possibilities when considering the need for better services and a big population. A poorer population and limited healthcare resources represent a problem as well as a benefit for entrepreneurs (Kalo et al., 2016). Emerging economies are characterized by underdeveloped markets and lack of strong institutions, which is reflected in weak laws, heavy bureaucracy, and institutional inefficiency (Khanna and Palepu, 2000). On the other hand, these 'institutional voids' or failures are seen by entrepreneurs as an opportunity to create businesses that promote social and economic development (Mair and Marti, 2009).

Digitalization not only improves productivity within the health sector, but also provides better results, greater quality and reliability, greater patient autonomy and higher quality of life. The main result is the provision of high-quality, patient-centred healthcare solutions at affordable costs and open to all (Bessant et al., 2017).

12.3.2   Data Collection

The two cases selected have in common that both are start-ups using digital technology to solve health and social problems. Data was collected by interviews and website research. Interviews were conducted with entrepreneurs and staff from the two Brazilian start-ups via semi-structured script, emails and WhatsApp texting. Table 12.1 summarizes the interviewees, their abbreviation for the text citations, dates and means of communication.

12.3.3   Data Analysis

The cases were analysed based on semi-structured interviews and documents (Flick, 2002; Simons, 2009). The combination of different sources of data aimed at improving the research validity. Empirical data were related to previous theoretical literature to give rise to discussions. The analysis started assessing categories at the innovation process of each firm individually. Afterwards, the cross-case analysis searched for similarities and for complementary knowledge between the two case studies.

The data were categorized according to the theoretical background in this study, based mainly on the categories of analysis that could respond the research question. They were: types of stakeholders participating in the development of the products; time in the innovation process when participations happened; the means through which participations happened; types of knowl-

Table 12.1   Interviewees

Respondent	Abbreviation	Details of interview
SOSPS		
Founder of the company and current commercial director	I1	Audio and text messages exchanged in August to November 2017 and January 2018
Founder of the company and hospital manager	I2	Email in January 2018
Founder of the company	I3	Email in January 2018
Patient and app user	I4	Email in January 2018
Patient and app user	I5	Email in January 2018
Patient and app user	I6	Email in January 2018
Hospital manager	I7	Email in January 2018
Hand Talk		
Founder of the company and current commercial director	I8	Audio and text messages exchanged in August, October, November 2017 and February 2018
Translator. She receives requests and suggestions from users	I9	Videoconference in October 2017
Marketing and responsible for feedback from external community	I10	Audio and text messages exchanged in October 2017

edge shared from stakeholders' participation; how firms assessed and used that knowledge; practical results from stakeholders' participation. However, as Graebner, Martin and Roundy (2012, p. 278) state, "the advantage of qualitative data for interpretive research is that researchers do not need to impose particular conceptual frames on the data prior to data collection". Therefore, we did not develop a rigid conceptual framework before the data collection, allowing the informants to express their experiences and establish causality mechanisms that could involve, for example, tensions and feedback loops.

12.4   CASE DESCRIPTION: SOSPS

Healthcare is one of the main social issues in developing countries. It is even more significant in large cities like São Paulo, which has a population of over 12 million. Given that the majority of people in Brazil still do not have full coverage of essential public health services, private hospitals and clinics need to step in mostly in emergency cases. Yet, it does not solve the problem of overcrowded hospitals and long medical waiting lists. Cases that are not so urgent represent 75 per cent of the medical appointments in an emergency hospital.

Two entrepreneurs from São Paulo were highly motivated by their own experience in the health sector where they worked as managers for many years. The app 'SOSPS' (in free translation SOS Emergency Hospital) has a function that allows to check the nearest hospital and its waiting time. It uses Manchester Protocol to classify cases of emergency (red) or very urgent (orange) in order to be able to direct them to the nearest hospital. Not so urgent cases are directed to hospitals with a shorter waiting time. Launched in January 2017, the app has been working with 35 clinics and hospitals in São Paulo and neighbouring cities and have around 10,000 users.

The waiting time data from some hospitals (between 35 and 40 per cent of the base) is captured by automated information using scrapers (software used to extract data and contents from websites). In the remaining base, a back-office team is needed to maintain daily interactions with hospitals in order to obtain the information.

12.4.1   Stakeholders' Participation in the Idea Generation Phase

Interviews with regular patients, on-call doctors, hospital managers, patients with chronic diseases (due to the greater recurrence of visits to hospitals) and the entrepreneurs' own experience were used to design the app. These interviews took place in two different moments: during the pre-project phase and the during the app development phase.

At the pre-project phase, users were able to contribute by evaluating the choices and by sharing their opinions on emergency room services as well as regarding the app's responsiveness to the waiting time. That included emergency room occasional users, parents of small children (paediatrics services) and chronic patients (greater recurrence of use). Doctors and hospital managers also contributed by identifying the challenges as well as the positive and negative opinions related to the receptivity to the product.

During the app development phase, users were able to contribute by testing the app's usability and information architecture. Doctors and hospital managers helped to define standards, minimum refresh periods, appropriate specialties for launching the least feasible product, services, screening protocols and essential information that should be made available to users.

12.4.2   Generating New Knowledge From Stakeholders' Participation

Users constantly send their suggestions and indications from hospitals to be monitored. The company replies to all of the feedback it receives. It tries to address the feedback received whenever it is possible. When the company is not able to address the problem, the app developers inform the users that they are working on the problem at hand and that users will be notified as soon as a solution is found. App developers, through formal and informal conversation and meetings, try to better understand users' problems to be able to develop features that increase the app value perception for healthcare professionals.

12.4.3   Applying New Knowledge and Adapting the Product

Although many users send their messages via Google Play or App Store, most do not return to these channels to check for answers or replies. In order to monitor and answer such users, the company adopts a proactive stance by searching users in other digital platforms to be able to both give feedback and better understand user's perception. In other words, the company is always screening and analysing information based on feasibility, sense of urgency and complexity.

12.5   CASE DESCRIPTION: HAND TALK

Seventy million people around the globe are deaf and struggle to communicate. Many deaf people have compromised development, since cognition is strongly related to language development. Brazil has more than two million deaf people, of which one million are under 19 years old. Very few of them are bilingual, being able to speak both Libras (Brazilian sign language) and Portuguese. In Brazil, the rights of deaf people to education and health as well as equal participation in the society are secured by legislation.

The inclusion of deaf people into the society is the focus of Hand Talk. It helps improve the independence of deaf people by increasing their interaction with the 'hearing' community, allowing deaf people to get a formal education and to join the job market.

The start-up works on two main platforms: a smartphone application (app) and a website plug-in. The app presents an animated avatar named Hugo who converts speech into sign language, acting as a personal interpreter for the deaf. Turning the audio into animations of gestures requires extremely detailed programming due to the precise nature of signing which includes facial expressions. To improve the clarity of the signing and demonstrate the precise signs, Hugo's hands and face are oversized.

The website translator is represented by an accessibility button on the screen. When enabled, Hugo automatically translates the selected texts into the sign language. These accessible websites can reach millions of people who did not have access to that content before. Thus, the deaf gain autonomy to obtain information and knowledge from the web. Although Hand Talk currently only translates Portuguese into Libras, the sign language used in Brazil, it has already been downloaded by 1.5 million people and will soon translate other languages.

12.5.1   Stakeholders Participation in the Idea Generation Phase

None of the three founders has a deaf family member. So when the founders began to outline the plans for the app, they ran three workshops with deaf people and their families to present the idea of the app and get feedbacks. Each workshop had between 10 and 25 people from the city of Maceio, where the start-up is located. The founder, interviewee I8, comments that this initiative was very important to get closer to users, learn about their daily challenges and understand their problems. From workshops, Hand Talk got inputs about how to help the deaf and validated the idea. The new knowledge helped the start-up to shape and design the product.

However, the reach of workshops was only local, not being able to cover users from different regions in Brazil. So the start-up needed to find another way to include external partners in the innovation process. Another challenge at this point was that interpreters – who would have a lot of useful information for the development of the innovation – did not help Hand Talk, because they thought that the product would take their jobs. "Disruption causes discomfort", says the founder. But soon interpreters understood the value of the app and the bigger mission behind the new product. Nowadays, interpreters are big partners for the improvement of the app.

12.5.2   Stakeholders' Participation after the Product was Launched

Currently, the participation of external partners happens in two ways: online and face-to-face. Most of the feedback from users comes from the app through the online Feedback area. The digital technology allows the start-up to have a broad reach of users, and it receives hundreds of messages each month.

Interviewee I10 is responsible for marketing activities and for receiving online feedback from the external community. She explains that she reads all feedback coming through the app and categorizes the ones suggesting improvements or changes. The founder (I8) comments that the team evaluates the recurrence of the feedback. If more people are asking for the same tool, its development will probably be prioritized. "This is how we adapt our products to be more useful for users", he explains. There is also a collaborative platform where interpreters all over Brazil help Hugo to learn sentence structures to translate better.

The second way to include stakeholders into the improvement of the products is via face-to-face contacts. The National Federation of Deaf is a partner of Hand Talk. Also, the team attends events all over Brazil, some of them are related to the deaf community, where there is a direct contact between Hand Talk and the deaf. When giving an interview for this research, the founder (I8) said that he had just visited an educational centre for the deaf in the State of São Paulo (Centro de Educação para Surdos Rio Branco). He wanted to better understand how Hand Talk could help children by observing classes. He explains:

> If we can make an impact on children's lives, the result is much greater, as they will have fewer losses throughout life. A deaf adult has already missed many opportunities, has been through school at a loss, and has a very small chance of having a decent job that would allow him/her the necessary financial return to support a family.

The interpreter at Hand Talk (I9) has a deaf brother, the reason why she decided to study Libras Interpretation as her undergraduate bachelor. She brings to Hand Talk the everyday experiences and challenges from the family of a deaf person.

12.5.3 Generating New Knowledge From Stakeholders' Participation

When the interviewee I10 reads all feedbacks coming from the app, she filters and categorizes them as: suggestions, missing signals and functioning problems. The founder (I8) comments that people communicate the same thing in different ways. That is why I10 needs to read *all* the messages.

After grouping them in sub-categories, she ranks the most requested signals and big fields of signals. The most requested ones go to the second evaluation: feasibility (time and complexity to be developed which gives also the estimated cost). If the suggestion is for a new signal, it is sent straight to the firm's interpreter. If it is a function error of the app, it goes straight to the technical department.

12.5.4 Applying New Knowledge and Adapting the Product

Several new signs or groups of signs are developed from users' suggestions. Also new features were developed from users' experience. In the beginning, the founders had planned the app to work as a translator for the deaf. Now, Hand Talk has an education function that teaches sign language because 75 per cent of users are hearing people who want to communicate with the deaf. Also the possibility of repeating translations and sharing translations by sending videos were developed from users' suggestions.

12.6   CROSS-CASE DISCUSSION

The generation of ideas of the two start-ups began from entrepreneurs. However, in order to be able to develop the new product, both start-ups had to establish a business model which included stakeholders. Both innovations were only successful because they managed to include external partners' knowledge on their innovation process, regarded as an important factor for RRI (Lubberink et al., 2017).

Concerning knowledge exchange on the idea generation, SOSPS had strong influence of tacit knowledge from entrepreneurs' own experiences as workers in the health sector. However, SOSPS does not include users in the improvement of the app after it has been launched. On the other hand, SOSPS depends on connections with partners in order to function. It depends on the information from hospitals on a daily basis. Different from SOSPS, Hand Talk does not need partners to function in terms of offering its services. Another difference is that, since its launching, Hand Talk counts on users' participation for its products' improvements. The start-up went through an adaptive process of entrepreneurship having to change the strategy of including users and getting feedback from them; instead of workshops with focus groups as in the beginning, now the firm focuses on online communications and informal face-to-face interactions.

From both examples and from literature of knowledge management (Hau et al., 2013; Nonaka and Takeuchi, 1995) and stakeholders' inclusiveness and responsiveness for RRI (Stilgoe, Owen and Macnaghten, 2013), we can say that the firms manage knowledge from stakeholders using separate structures. The inclusion of stakeholders may generate inputs for the development of a new product in the shape of tacit and explicit knowledge. The two companies established different routines to deal with the different type of knowledge coming from stakeholders: participation of external agents happen via different structures, and the internal use of this external knowledge is also managed in a different way.

As a result of this process, we have RRI created from stakeholder's knowledge inclusion. The two processes of knowledge management are:

1. The process regarding face-to-face interactions generates sharing of tacit knowledge. It can happen, for example, via focus groups, informal meetings with stakeholders and visits to users' sites. The outcome is the generation of understanding about long-term problems and users' needs on a daily basis.
2. The process starting from online interactions generates explicit knowledge from a broad network. It can happen, for example, through the firms' mobile applications, websites and social media. The outcome is the gener-

ation of a systematic registration of suggestions for improvements and new features. The key functions in this phase are processing and combination of knowledge.

12.7   EMPIRICAL IMPLICATIONS

Given the Brazilian context and its health system already described, techno-logical solutions have been developed focused on improving the quality of healthcare and quality of life. The two case studies addressed by this study offer us some key points. First, both have shown potential for scaling-up their operations. Both apps were developed to tackle problems in contexts involving a large proportion of the population. Despite having different scopes, both applications seek to improve users' quality of life. In short, they combine a socially responsible mission with a market-oriented approach. Second, both companies have established the participation of different stakeholders since the firms' conceptions. Stakeholders are encouraged to actively participate with feedbacks and insights from the developing stage to the usage phase. By doing this, both companies are able to use the knowledge generated through stakeholder participation to improve and develop more efficient products.

Finally, stakeholder participation is not limited to final users. By allowing actors such as the National Federation of Deaf, on-call doctors and hospital managers, entrepreneurs are able to show the benefits of the app resulting from stakeholder participation. Even when facing divergent opinions, as in the case of Hand Talk interpreters, the entrepreneurs did not stop encouraging stakeholders' participation.

12.8   CONCLUSION

This chapter aimed to analyse how start-ups manage stakeholder participation in the RRI context. We presented two case studies of Brazilian start-ups that develop online solutions for social and health problems. Due to its scale and importance, the healthcare sector offers business opportunities to many companies. The two start-ups in this chapter reinforce that firms can become economically sustainable when developing e-healthcare products.

The analysis of the case studies showed that the inclusion of stakeholders occurs through the exchange of tacit and explicit knowledge. Companies need to develop different routines, processes and structures to deal with different types of knowledge and increase the result obtained through inclusion.

There are three main limitations associated with the results. First, the small number of case studies from the health sector does not allow the validation of results to a wider population of start-ups. The second limitation concerns the context of an emerging market. Despite the common characteristics of these markets, some attributes may be unique to Brazil. Researching companies in other emerging markets may bring new insights. Third, the case studies here are young start-ups. Consolidated companies, with larger structures, may have developed different processes, routines and structures, as well as obtained different results from stakeholder participation.

REFERENCES

Agência Nacional de Saúde Suplementar (ANS) (2018). Taxa de cobertura (%) por planos privados de saúde (Brasil – 2008–2018), 4 August. Retrieved 26 March 2018 from http://www.ans.gov.br/perfil-do-setor/dados-gerais.

Associação Nacional de Hospitais Privados (ANAHP) (2018). Disponibilidade de leitos no país está aquém do índice da OMS, 4 August. Retrieved 24 March 2018 from http://anahp.com.br/sala-de-imprensa/disponibilidade-de -leitos-no-pais-esta-aquem-do-indice-da-oms.

Becerra, M., Lunnan, R. & Huemer, L. (2008). Trustworthiness, risk, and the transfer of tacit and explicit knowledge between alliance partners. *Journal of Management Studies*, 45(4), 691–713.

Bessant, J., Alexander, A., Wynne, D. & Trifilova, A. (2017). Responsible innovation in healthcare: the case of health information TV. *International Journal of Innovation Management*, 21(8), 1740012.

Burget, M., Bardone, E. & Pedaste, M. (2017). Definitions and conceptual dimensions of responsible research and innovation: a literature review. *Science and Engineering Ethics*, 23(1), 1–19.

Constituição da República Federativa do Brasil de 1988 (2001). [Coleção Saraiva de Legislação] (21a ed.). São Paulo: Saraiva.

De Silva, M., Howells, J. & Meyer, M. (2017). Innovation intermediaries and collaboration: knowledge-based practices and internal value creation. *Research Policy*. doi: 10.1016/j.respol.2017.09.011.

De Souza, R. (2002). O sistema público de saúde brasileiro. Ministério da Saúde. Retrieved 22 March 2018 from http://www.inesul.edu.br/site/ documentos/sistema_publico_brasileiro.pdf.

Demers-Payette, O., Lehoux, P. & Daudelin, G. (2016). Responsible research and innovation: a productive model for the future of medical innovation. *Journal of Responsible Innovation*, 3(3), 188–208.

Flick, U. (2002). Qualitative research–State of the art. *Social Science Information*, 41(1), 5–24.

Graebner, M. E., Martin, J. A. & Roundy, P. T. (2012). Qualitative data: cooking without a recipe. *Strategic Organization*, 10(3), 276–284.

Hau, Y. S., Kim, B., Lee, H. & Kim, Y.-G. (2013). The effects of individual motivations and social capital on employees' tacit and explicit knowledge sharing intentions. *International Journal of Information Management*, 33(2), 356–366.

Instituto Brasileiro de Geografia e Estatística (IBGE) (2018). Projeção da população do Brasil e das Unidades da Federação, 4 August. Retrieved 26 March 2018 from https://www.ibge.gov.br/apps/populacao/projecao/.

Jasimuddin, S. M., Klein, J. H. & Connell, C. (2005). The paradox of using tacit and explicit knowledge: strategies to face dilemmas. *Management Decision*, 43(1), 102–112.

Kalo, Z., Gheorghe, A., Huic, M., Csanadi, M. & Kristensen, F. B. (2016). HTA implementation roadmap in Central and Eastern European countries. *Health Economics*, 25, 179–192.

Khanna, T. & Palepu, K. (2000). Is group affiliation profitable in emerging markets? An analysis of diversified Indian business groups. *The Journal of Finance*, 55(2), 867–891.

Leonard, D. & Sensiper, S. (1998). The role of tacit knowledge in group innovation. *California Management Review*, 40(3), 112–132.

Lettice, F., Rogers, H., Yaghmaei, E. & Pawar, K. S. (2017). Responsible research and innovation revisited: aligning product development processes with the corporate responsibility agenda. In Brem, A. and Viardot, E. (eds), *Revolution of Innovation Management* (pp. 247–269). London: Palgrave Macmillan.

Lubberink, R., Blok, V., Ophem, J. v. & Omta, O. (2017). Lessons for responsible innovation in the business context: a systematic literature review of responsible, social and sustainable innovation practices. *Sustainability*, 9(5), 721. doi: 10.3390/su9050721.

Mair, J. & Marti, I. (2009). Entrepreneurship in and around institutional voids: a case study from Bangladesh. *Journal of Business Venturing*, 24(5), 419–435.

Nonaka, I. & Takeuchi, H. (1995). *The Knowledge-Creating Company: How Japanese Companies Create the Dynamics of Innovation*. Oxford: Oxford University Press.

Nonaka, I., von Krogh, G. & Voelpel, S. (2006). Organizational knowledge creation theory: evolutionary paths and future advances. *Organization Studies*, 27(8), 1179–1208.

Ramamurti, R. (2012). Competing with emerging market multinationals. *Business Horizons*, 55(3), 241–249.

Simons, H. (2009). *Case Study Research in Practice*. London: Sage publications.

Smith, E. A. (2001). The role of tacit and explicit knowledge in the workplace. *Journal of Knowledge Management*, 5(4), 311–321.

Stilgoe, J., Owen, R. & Macnaghten, P. (2013). Developing a framework for responsible innovation. *Research Policy*, 42(9), 1568–1580.

TelessaúdeRS (2018). Institutional presentation, 4 August.

von Krogh, G., Ichijo, K. & Nonaka, I. (2000). *Enabling Knowledge Creation: How to Unlock the Mystery of Tacit Knowledge and Release the Power of Innovation*. Oxford: Oxford University Press on Demand.

Yenicioglu, B. & Suerdem, A. (2015). Participatory new product development: a framework for deliberately collaborative and continuous innovation design. *Procedia – Social and Behavioral Sciences*, 195, 1443–1452.

13. Responsible innovation and commercialisation in the university context: a case study of an academic entrepreneur in digital healthcare

Bernard Naughton and Lene Foss

13.1 ACADEMIC ENTREPRENEURSHIP AND RESPONSIBLE INNOVATION

Some of the most lucrative cases of research commercialisation in academia relate to the development of medicines. However, this is not always responsible innovation as these medicines are often too expensive for low and middle-income countries (Siegel and Wright, 2015). It is important for academics involved in the commercialisation of research to bear in mind the needs of society and strive to innovate in a responsible way. Universities are increasingly encouraged to take an 'entrepreneurial turn' by licensing new technology, creating university spin-offs and expanding university–industry relations (Foss and Gibson, 2015a). Another way that universities can be entrepreneurial is to pursue *academic entrepreneurship* by commercialising knowledge and research findings (Klofsten and Jones-Evans, 2000; Roessner et al., 2013). One would expect that all publicly funded universities innovate responsibly. However, there is little research which evaluates the level of responsibility within academic entrepreneurship activity. Research suggests that universities need several capabilities to facilitate the venture-formation which includes creating new paths of action, balancing both academic and commercial interests and integrating new resources (Rasmussen and Borch, 2010) without any clear direction to be responsible in their innovation. Research also shows that the opportunity recognition capacity of the academic and prior entrepreneurial experience are the most important predictors of academic entrepreneurship, whereas the activities of Technology Transfer Offices (TTOs) play a marginal and indirect role in driving academics to start new ventures (Clarysse, Tartari and Salter, 2011) with little information which identifies the impact of responsible innovation on commercialisation.

One study indicates that the entrepreneurial activities of the scientists heavily depend on patenting activity, entrepreneurial experience, personal opinions about the benefits of commercialising research and close personal ties to industry (Krabel and Mueller, 2009) but it is unclear if these personal opinions always come from a responsible perspective. Interestingly, there seems to be limited empirical studies of the 'culture' dimension (institutional, departmental, and individual attitudes and norms) in the entrepreneurial architecture framework developed by Nelles and Vorley (2010). In their conclusion of the entrepreneurial architecture of 10 universities in five countries, Foss and Gibson (2015b) argue future studies to investigate the cultural dimension of academic entrepreneurship are required. This culture investigation may include how responsible innovation is seen and understood in the academic context and whether systems are in place to ensure RI remains at the core of research and innovation.

13.2   THE EMPIRICAL CONTEXT

13.2.1   UK NHS Digitalisation

The National Health Service (NHS) in the UK offers free healthcare to UK citizens at the point of access. The NHS has been digitalising for many decades. However, the digitalisation of the UK NHS has not been a complete success. An example of poor digitalisation in the NHS is the National Program for IT (NPfit) project. In 1998 the NHS Executive set a target for digital patient records to be in place across all NHS Hospital trusts by 2005. By 2002 only 3 per cent of NHS trusts were likely to reach this figure (Hendy et al., 2005). Fast forward to 2015 and the NHS announced that patient records will be electronic by 2020. This is an overview of a complex implementation issue; however, it is clear that the NHS has failed to reach their digitalisation targets. There were a number of reasons for failure to meet the initial 2005 target. These reasons are technical, financial and sociocultural (Hendy et al., 2005). In the study by Hendy et al. in 2005 there was a lack of stakeholder involvement which hindered the NPfit project. The lack of stakeholder engagement associated with the failure of the NPfit project strengthens the argument for an RI approach in digital healthcare project management. An RI framework could be used systematically in all UK digital healthcare projects to ensure that the failures, similar to the NPfit project, are not repeated.

13.2.2   UK Academic Policy Context

The UK has a history of encouraging UK universities to engage with NHS hospitals and the private sector to demonstrate impact from their research. In recent years there has been the Lambert review (Lambert, 2003) which identifies that some universities collaborate well with industry, but there is more that can be done to increase collaboration. More recently there have been policy moves including the development of the UK Research and Innovate (UKRI) body which has a budget of 6 billion and is used to fund the "Partnership of universities, research organisations, businesses, charities, and government to create the best possible environment for research and innovation to flourish" (UK Research and Innovation, n.d. (b)). This new body recognises that successful collaborations are necessary for generating meaningful impact. Furthermore, the UK government considers research impact as a high priority when examining university research outputs. The UK government performs the Research Excellence Framework to assess the excellence of research generated from UK universities. Impact case studies play a big role in this framework which gives universities the opportunity to showcase the impact that their research has had on society. Impact can be demonstrated through the inclusion of research papers in government policy, qualitative feedback from practitioners and collaborations or consultancy between academia and private business.

Digitalisation is an important part of UK healthcare. Digitalisation can improve efficiency and performance and can save space through the digitalisation of paper processes. The current trend to digitalise within healthcare, coupled with the UK government's continued push for academics to engage with businesses will present opportunities for UK academics with research experience in digital healthcare. However, some academics may not be equipped with the necessary skills and attributes to collaborate with the private sector effectively and those that do will face obstacles to successful collaboration and impact generation.

13.3   FRAMEWORK AND METHODOLOGY

To explore our research question theoretically we use the RI framework by Stilgoe, Owen and Macnaghten (2013). They propose four dimensions of responsibility: anticipation, inclusivity, reflexivity, and responsibility, see Box 13.1.

BOX 13.1 FOUR DIMENSIONS OF RESPONSIBLE INNOVATION

1. Anticipation: This involves looking forward, horizon scanning or using foresight.
2. Inclusivity: User-centred design and including stakeholders through focus groups and disseminating findings in an open and easily accessible way.
3. Reflexivity: Rethinking and redefining based on the views of stakeholders
4. Responsiveness: This follows inclusivity and reflexivity and relates to how the entrepreneur responds to the opinions of their stakeholders.

In applying these dimensions in an academic entrepreneur setting within a university, we developed an interview guide (Appendix 13.1). As there are few close-up studies of academics pursuing responsibility in commercialising science within a university context, this case study takes a narrative approach (Flyvbjerg, 2006). Thus we performed a qualitative interview with this professor which records his story of how he innovates within a university context. Although our study is not an ethnographic study, we are influenced by the term Denscombe (1998, pp. 68–69) uses "understanding things from the point of view of those involved rather than explaining things from the outsiders' point of view". Consequently, our interview is conversational. We aimed at letting the interviewee, hereafter called Alan, tell us his honest experiences of how he performed digital research within the university structures and context. The interview was transcribed and thematically analysed. In the next section, the narrative of the professor, we have framed the story along some important lines of demarcation that came up in the interview. Most of them are a reflection on the themes in our interview guide, whereas others came up as new elements that situate the professor's effort to execute RI in a larger framework in a university context, that is, psychological contracts, culture for commercialisation, other academics' perception of commercialisation and so on.

13.4 THE STORY OF ALAN

Alan responded positively to the interview; he appeared honest and forthcoming. He was passionate and excited about his work and was pleased to speak about himself, his journey and his achievements. Alan is in his early sixties, and began his career as a scientist in the pharmaceutical industry. He worked in sales, marketing, R&D and international manufacturing in factories in Sweden, Paris and the UK. After ten years in this industry and for ethical reasons he moved, to academia. He articulated this decision in this way:

> Sales and marketing finished me off. I just couldn't ... couldn't do it; I just couldn't go and make a case for something I didn't believe in.

In academia, he was first taken on at a mid-sized UK university, and was initially employed on a one-year rolling contract which was renewed on the basis of continued funding generation. One of his first academic research projects related to the digitalisation of healthcare records. In this scenario, he aimed to digitalise a number of processes within healthcare and began with the digitalisation of medical paperwork required for reimbursement. He then moved on to research decision support tools, then on to avatars for education and training and then investigated immersive interactive learning tools. His success has been illustrated through successful Research Excellence Framework impact case studies (a case study report explaining the measurable impact of research on society), collaborations with hospitals, national bodies and the generation of over 30 million pounds for his research institution through the commercialisation of digital projects. However, Alan did not have a traditional start to academia. He was incentivised to innovate to maintain his position. He explains that he was brought in as "a completely unknown quantity" to do a specific project. The university was happy with his work, and in his words they said:

> Right, we'll give you a year, and if you earn enough money to pay for your salary by the end of the year, we'll give you another year.

And although his employment rights will have changed dramatically; in theory, his written contract is the same. A psychological element of this contract exists which means that considering Alan can generate income, his position is tenable:

> Bring enough money in. You have a job. Don't bring in the money. You don't have a job. Straightforward as that! So, I've always brought in money.

The interview reveals that Alan is a successful academic entrepreneur with soft skills such as relationship- and trust-building. He relies on his soft skills and explains that to get his way he would tread carefully around colleagues required for projects to progress and in his words, he would "give the occasional stroke" to maintain relationships. He also explains that it takes time to build the relationships necessary to succeed within the university. He identifies trust and relationship building as important in commercialisation within a university:

> It goes back to Aristotle ... if you want to influence a man, first convince him you are his true friend.

Many UK universities and private companies have formed successful partnerships with teaching hospitals; in some cases for digitalisation projects. However, it is clear that the NHS has a history of failed digitalisation projects, as seen in the NPfit project mentioned previously. Alan has developed an effective strategy for succeeding in the university context and has generated much of his research income through NHS-funded projects. The next section reveals his struggles and successes and we aim to understand the part that RI has played in these outcomes.

Alan has many years of experience conducting digitalisation projects within the NHS and private healthcare sector. This digitalisation includes communication digitalisation, replacing paper processes, computerised decision support tools and avatars for healthcare professional and patient education. Alan faced funding hurdles in the early years and established that the cost of using large external companies to create digital solutions was unsustainably expensive and that the affordable solutions were as he described "Clunky". As he experienced barriers along his journey there was a need for university funds to maintain his drive for digitalisation. Alan took it upon himself to approach the vice chancellor for funding for a computer programmer and two graphic animators to fund a digital avatar project, describing this approach as, "The hardest thing I've probably had to do since I've worked for the university". It happened that the vice chancellor funding which was aimed at creating technology for internal teaching use, was then expanded to provide a variety of digital products. These products brought in further revenue from the NHS and private healthcare firms and allowed his team and technologies to expand to include immersive technology. The professor's digital avatar and immersive technology project then became a university centerpiece:

> What happened then was because this was very impressive and very new, said vice-chancellor, every time he had a visitor, this would be trotted down; nothing else to show them. So, it started generating outside interest.

Alan aspired to facilitate wider access to his digital solutions. He understood that as an expensive university centrepiece, this digital technology would struggle to impact large patient numbers and the professor investigated how this technology could be miniaturised into mobile phone apps. His latest app has been approved by a national body and has served thousands of users.

13.4.1 The Value of Psychological Contracts

Alan's story reveals the role of psychological contracts in academia. A psychological contract refers to the unwritten set of expectations between the employer and the employee. (Rousseau, 1989). In the UK a typical university professor has a full-time contract that is difficult to terminate without evidence of gross misconduct. The psychological contract between a professor and a university plays a lesser role than the written agreement, due to robust job security. In private employment, a psychological contract can often play a more influential role. In a private company, the employee and employer understand that poor performance can result in the termination of employment or the unlikeliness of promotion.

The strong job security associated with a tenured professor position can result in apathy and complacency. Alan does not have such an arrangement. Although Alan's employment rights will have changed drastically since his initial employment, he is still technically on that same written one-year rolling contract and is influenced by a psychological contract which is more akin to one seen between a private company and an employee. The psychological contract ensures he generates income in exchange for a renewed written contract and appears to have encouraged this entrepreneurial professor to succeed; which is evidenced by his generation of millions in revenue for the university:

> The last count I think I made, probably over my career £32 million for the university. Something like that. It's not a huge amount, but I very rarely make them less than £1 million per year.

The psychological contract between Alan and the university began when he was first employed. Trust began to develop as he consistently generated £1 million a year for his institution. This psychological contract with the university then evolved as he generated further funding and was eventually granted professor status. Further psychological contract evolution is evidenced when this professor approached the vice chancellor for digitalisation funding to push his research forward. The relationship has stabilised over many years, likely due to evidence of pedigree, continuing success and trust. It takes an individual and university with entrepreneurial attributes to agree to such an employment arrangement and takes a professor with resilient qualities to maintain the psychological contract needed to sustain this academic post for so many years. It is often the case with psychological contracts that they are breached more often than not (Robinson and Rousseau, 1994), however in this case there was no mention of breach which facilitated the success of this relationship.

The presence of a psychological contract may also play an important role in RI. The psychological contract involves an agreed understanding of what is expected from both parties. A healthy psychological contract between a professor and a university that encourages RI, may facilitate a professor to conduct RI practice, due to mutual respect and the understanding that failure to deliver on the psychological contract could result in the termination of position. Furthermore, the transfer of RI practice behaviour is likely to cascade not only from the university to the professor, but also to his researchers and students. Without the presence of a strong psychological contract, the professor has less incentive to follow university standards, values, and practices.

We suggest or conclude that Alan's entrepreneurial attributes and the trusting psychological contract in place between him and the university drove Alan to innovate in digital healthcare continuously in a responsible way. "As long as I stay employed – just about! – I make the surplus, it keeps the unit going; it keeps me having an interesting life." This quote alludes to the concept that the practice of RI demonstrated in this case study also played a part in maintaining this success.

13.4.2 Barriers and Success

Alan experienced some successes including product commercialisation in early digitalisation projects, resulting in a spin-out company which now turns over between £400,000 and £500,000 a year. During interviews, Alan identified his most recent innovation, a decision support tool, which has had 15,000 sessions by 11,000 individual users and has been endorsed by a national body. Alan describes this as:

> Probably the most successful one we've done to date.

It appears that Alan puts more emphasis on impact than financial gain. The app has not generated income, but is still described as being more successful than his profit-making ventures, due to its wide use and endorsement by a national body. The need to generate funding to maintain Alan's contract seems to have driven him to create high levels of research income. Despite some success and the commercialisation of a spin-out company, there have been commercialisation barriers relating to innovation within the university context such as the lack of sufficient marketing and commercialisation support:

> If I wanted to commercialise it further I've got a real challenge because it's a small company and I'm an academic; I'm not a marketer. So, if I want to sell it I need salesmen, I need a salesforce, I need some sort of marketing platform. If I want a marketing platform, I need a budget for marketing. No chance of getting that inside the university. None whatsoever. That's just not how they see the role. The

university loves the idea of you doing knowledge transfer and bringing in income as long as they could do it without them having to spend any money, basically, is the message.

Alan then makes clear that he is working on an emerging technology, and that he is not sure how to commercialise it. He reveals it as being "So big that I don't have the energy to do it". He also describes that getting a share of the market voice and competing with large multinational companies is difficult as a small group or organisation. Alan clarifies that conservative procurement strategies also rule out small start-ups and that some firms will only buy from companies with three years' trading accounts which automatically disbars young start-up companies.

Alan tells us that the revenue generated from intellectual property (IP) poses another barrier to commercialisation. The type of IP contract in his institution determines the percentage of IP profits retained by the entrepreneur. A low percentage of IP profits deters academics from innovating and make commercialisation less worthwhile. In his institution, the majority of income generated from an entrepreneur's IP is retained by the university. It is apparently down to particular contracts but virtually 50 per cent of the commercialisation funding goes to the central university, the remaining 50 per cent goes to the school. If there is intellectual property, the inventors receive a percentage of the funds remaining. Alan spells out "The inventors get a percentage of the stuff that's left over, but you don't get very much at all because you get taxed twice". He explains the hypothetical situation of generating £100,000 from IP. "If you get £100,000 worth of income, the first thing they do is they'll take £20,000 to pay the corporation tax. So, then that becomes £80,000. The £80,000 is then split in two. So, £40,000 goes to the university and £40,000 goes to the school. That is split in two again so £20,000 goes to the school and £20,000 goes to the inventor. Then, you get income tax and national insurance, etc., so if you're lucky you'll come away with somewhere between £8,000/9,000 from £100,000 worth of income generation." Sometimes, the university will say "no, that's not intellectual property" which means the inventor gets little or no income from the commercialisation of innovation. Alan also explains that there is a low threshold for what users will pay regarding digital healthcare solutions. He reveals that healthcare networks are always pleased he produces a product if it gets to them free of charge. However, if he approached healthcare institutions and asked if they would you like to buy a tool for £2 each, the appetite would not be there. "The threshold is very low, particularly in healthcare in the UK". Arguably, there seems to be an unfair remuneration for academics who generate IP and healthcare facilities have a reduced appetite for digital healthcare solutions that they need to pay for. The culmination of these two factors act to disincentivise academics to commercialise digital healthcare technologies.

13.4.3   Technology Transfer Offices

The entrepreneur later mentions the limitations of innovation transfer offices. He explains that his university does not have a technology transfer office but has another department which deals with university innovation. He explains that it is supposed to support academics with legal contracts and so on, but they won't proactively reach out to academics with projects or consultancy opportunities. Alan reveals that the transfer office staff do not necessarily understand how research and innovation work, especially regarding the implications of missing deadlines. This lack of infrastructure makes commercialisation even more difficult, as the academic must commit precious time in commercialising products without centralised support. The issues that Alan identifies illustrate a lack of RI principles within TTOs as TTO employees fail to understand the context and needs of key stakeholders, that is, academics and funders.

13.4.4   Academic Commercialisation Consensus

Alan also exposes in the interview that the perception of other academics is another significant barrier to commercialisation. Some academics believe that commercialisation should be conducted by private companies and not by academics: "We shouldn't dirty our hands with commercialisation" and that academics should "Sit here and pontificate and think great thoughts and teach our students and leave others to do that (commercialisation)". He acknowledges that culture is a significant challenge to commercialisation within universities. Alan describes that this outlook seems to be in contrast to the director of external engagement within the university. Alan paraphrases for the director of external engagement and says that the director thinks that academics should commercialise and innovate, but shouldn't be incentivised to do so and that it is part of their job:

> That attitude is "why should academics get anything extra? Surely we pay them for doing this?" The answer is no you don't; you pay us to teach and do research. You do this over and above. It takes a huge effort to go outside your comfort zone and go and negotiate with a major pharmaceutical company like company X.

The UK has historically tried to develop effective strategies to encourage academics to work closer with industry and it seems that the UK is continuing to encourage commercialisation and industry collaboration. Considering Alan's experience, it appears that some academics are still opposed to this way of working. This opposition may be due to a fear of the unknown, little financial or IP incentives to commercialise or the consensus that academics should pontificate and not commercialise. If a peer is fundamentally opposed to commercialisation, it is difficult to see the benefit of involving them in stakeholder discussions, no matter their suitability. Therefore, peer opposition to commercialisation may be a barrier to inclusive stakeholder engagement, which is argued to be the raison d'être of RI.

13.4.5 Knowledge and Perception of RI

When asked about his thoughts on RI Alan shared that sometimes there is pressure associated with private company-funded research. And that he "Can certainly see the logic in having innovation separated from vested commercial interests when it's for the benefit of the herd, public (the general public that follow accepted norms within their context)." He identifies that research funded by government or charities carries fewer obligations. Alan is engaged with the EU's major research programme Horizon 2020 and he is now trying to move away from private company-funded projects and work more with government, charity or foundation-funded research, describing them as "More sustainable for my team", even after years of industry collaborations. We all have a role to play in the promotion of RI, but in practice it is likely that the entrepreneur has most influence over the level of RI in place, and it is up to them to involve the right stakeholders in product development and manage the tensions which exist between industry funders and patient needs.

13.4.6 Inclusivity – Including Stakeholders

Alan communicates how he performs inclusive research. His first task is to establish a stakeholder group which includes a lead clinician, and some clinical area specialists. If the project involved primary care, he would get a GP and a clinical pharmacist and technical experts, animation team and they sit together and identify what the set of needs are. Alan explains that his diverse group helps him to make a decision. "My team is formed of clinical pharmacists, medics, epidemiologists, data analysts. I've had sociologists, psychologists, health economists, you know, whatever I need, actually, to help me make a decision." When we ask him how he uses the end user in this process, Alan makes clear that he involves end users throughout the innovation:

> So, you probably have one big meeting, a working group afterwards, then some work; quite a lot of email correspondence, so prototypes going backwards and forwards and then, probably, another meeting as we approach beta testing. Then we'll get a beta test prototype and we'll circulate it out to … we'll pilot it with a number if

it's going to healthcare professionals with a number of healthcare professionals and ask them to feedback as well.

We observe that through stakeholder involvement, he appears to participate in inclusivity which is a marker for RI. He appears to develop products with the right mix of experts and doesn't appear to exclude useful representatives including the end user.

13.4.7   Reflexivity – Rethinking and Redefining

Alan explains that he tests his digital prototypes with stakeholders, in order to gather or receive feedback. His standard format is to have one large working group meeting, followed by plenty of email correspondences, and prototypes going back and forth between himself and the stakeholders. This is then followed by another meeting and a beta test prototype which is piloted with some healthcare professionals where feedback is gathered. In including stakeholders, reflecting and acting on that input demonstrates a reflexive approach to innovation. There is an end user right at the beginning at the initial working group, who is also part of one of the small task and finish task groups who review the final product. Furthermore, Alan has some end users who will form part of the pilot and beta testing before the innovation is rolled out.

Interestingly, Alan reveals that a balance has to be made between fulfilling the needs of the end user and, the requirements of the payer. This separation in healthcare innovation means that sometimes the stakeholder wants one thing, but the payer wants another. This trade-off is described as creating "some tensions, and that's where we come in because we act as referee". When the payer is a pharmaceutical company, "you must be wise to their agenda". His technologies are not based on scientific findings alone but also based on practice-based advice. He expresses that "you are tempering the evidence base with the reality of clinical practice" demonstrates, in our view, that Alan demonstrates rethinking and redefining during product design.

From Alan's experience and research dealings in the pharmaceutical industry, he illustrates that there is a lack of reflexivity within large companies which can hinder innovation. Section managers of pharmaceutical companies often look after the interests of their respective area, but not the entire company as a whole. They don't necessarily consider how one product could be useful for a colleague in another department or whether the research investment will help the wider company. The section managers are more concerned with reaching their own goals and targets. This lack of reflexivity is in contrast to his approach where he reflects and acts on stakeholder feedback. Alan reiterates that large multinational companies do not often practice reflexivity and often fail to take on board the advice of those in practice and instead persist with the companies' agenda. Research with a company that is not reflexive can be an issue if that company pushes an agenda which is not in the best interest of society, or not what the stakeholder wants:

> One has to be terribly careful not to get embroiled in something that later on you would be embarrassed to be associated with.

Reflexivity has its positives if performed with the right partners. Alan informs the interviewer about a situation where reflexivity causes serious issues. He recently experienced an incident where he discussed his product with a pharmaceutical company and agreed to create a bespoke version of that product for the company. He had his work stolen before the company signed the licence agreement as promised. They started to develop the technology themselves. We interpret this as the dark side of reflexivity; engaging with partners that cannot be trusted. The dark side of reflexivity is probably the single most important obstacle to reflexivity in RI practice. Not only does IP theft impact the affected project, but it can have lasting effects on an innovators' willingness to collaborate with businesses, which has future knock-on effects in terms of funding and impact generation within a university. In this case study, Alan demonstrates a very professional and considered approach and instead of being slanderous he understands the need to maintain a relationship and the pros and cons of his potential actions:

> I guess the other lesson from that is you have to keep working with them. There's no point, then, throwing your toys out the pram. And gnashing and calling them names and writing off offensive letters or going to the *Sunday Times*, or anything like that. All of which I thought about! Because lose a battle, win a war.

Alan explains that maintaining a relationship despite deceit is important in the long run when researching with private firms. He also explains that large companies try to take ideas and IP from academia knowing that small and mid-sized universities do not have the funding or legal capacity to pursue litigation cases.

13.4.8 Responsiveness: Responding to Stakeholders' Needs

Through stakeholder involvement, Alan has been able to gain further insight and implement contextually appropriate suggestions to improve his technology. One example is the developing a children's app. Only through arranging a focus group with children did he realise the real questions that this app needed to address. One child asked; "You keep saying there are all these medicines that are really good …but what about the side-effects." The professor explained; "Of course there are some side effects." He realised that using statistics to explain the prevalence of side-effects was not suitable for children. So, he came up with this idea of explaining side-effects by using the digital metaphor of an event in a large venue, which he later embedded into his app. The digital metaphor shows 100,000 attendants in a venue, this then shows five people leaving the venue due to a side-effect after taking a drug. This visual approach explains that the chances of a child feeling a lot better are very high and the chances of a child having a side-effect, that is, leaving the venue, are very low. Through stakeholder engagement, this professor has produced a digital solution which matched the needs of the users.

Alan has also adapted his approach to suit clinical practice in different countries. In the UK, many healthcare facilities have consulting rooms. The app developed by the professor had created a simulated healthcare consulting room as part of this app. However, through responsive stakeholder engagement he realised that healthcare consultation rooms were not routinely available in all European healthcare facilities. He, therefore, had to re-model his approach. This example of responsive stakeholder engagement has ensured that his digital healthcare tools are suitable for each context which they are intended for and meet the needs of its stakeholders:

> In Germany, they don't do that. They didn't think consulting rooms were necessary at all. So, we had to redo a version for Germany without a consulting room.

Although responsiveness is an important part of RI, it can also act as a barrier. When the end user wants something different to either the payer or the creator of the digital solution it can cause some issues. "It creates some tensions". Responsive behaviour inevitably requires good negotiation skills to balance the wishes of all stakeholders. Some of the issues relating to responsive behaviour are augmented through working with larger firms and negated through working with smaller firms. "A small company can look at the value for the company as a whole. Whereas, a big company, a product manager, will look at what it means for him or her." Alan appears responsive and takes stakeholder opinion on board. He acknowledges, however, that lack of finance is an inhibitor of responsiveness. Through engaging with stakeholders, he can be overcome with innovative suggestions, but without the appropriate budget, he must decline some of these suggestions:

> You always take it on board. So, you listen. Whether you do anything about it is dependent on what your capacity is to make that change. But, would I ever be able to get the budget for that? Not a snowball's chance in hell.

We conclude that stakeholder engagement seems important in this case, and that it improves the design of digital products, making these products more relevant to users. Collaborating with smaller companies has its benefits, but despite the size of the company, it is challenging to balance the needs of the payer, the regulator and the user, especially in the absence of adequate funding.

13.4.9 Anticipation – Horizon Scanning

Being responsible means looking forward as well as being prepared for the future. This narrative has elements that evidences responsible working. Alan explained that to achieve his digitalisation goals he would require university funding to develop a sustainable digital agenda. Alan had the foresight to approach his vice chancellor to secure funding which would generate future commercial funding for his other activities. Alan also realised that smartphones were increasing in popularity and that there was a need to miniaturise his technology to facilitate widespread use. Alan also foresaw the need for two rather than one software developer and therefore employed staff with experience of app development. Although these pieces of evidence demonstrate that Alan looks forward, overall, anticipation appears to be less important to this professor that being inclusive, reflexive and responsive. This may be due to his busy schedule which means that he has capacity for dealing only with what he faces. By his own admissions he takes too much work on, which would explain why he lacks the time to be proactive and anticipatory in planning ahead too stringently. Furthermore, his written recurring one-year contract with the university would have historically prohibited activity which was too far in the future. His employment rights will have changed and his current position may be secure, however his previous arrangement may have conditioned him to place less priority on anticipation, which is a negative aspect of short-term academic contracts.

13.4.10 Ethical Activity

This professor has a rather negative opinion of collaborations with private companies, which within his context would largely be pharmaceutical companies. He explains that private companies want to 'maximise profits' and 'minimise the risk' (to themselves) he then explains that there is always going to be a disparity between stakeholder and shareholder needs:

> So, there's always going to be a tension between the needs of the patients and the needs of the corporation. The corporation's first responsibility is to their shareholders.

In contrast, Alan does prioritise the needs of the patient and sees himself as the middleman caught between the private company and the patient. He sees himself as a protector of patients which is a necessary and important role if pharmaceutical companies are out for themselves, as he describes. As the UK government encourages academics to conduct research with private companies, through incentives like the UKRI industrial strategy challenge fund we believe that it will be important to upskill academics to understand the risks and benefits associated with such collaborations (UK Research and Innovation, n.d. (a)):

> Whereas my responsibility is to the patients and that's how I see myself as one of the guardians or gatekeepers in that role to make sure that we don't do anything that we wouldn't think would be of benefit to patients. And we do turn down projects.

This narrative portrays a responsible entrepreneur who puts patient needs ahead of financial gains. Alan had the opportunity to receive £1 million in a research grant to conduct drug research, but when he proposed the project to his team they refuted the project, as they believed the dose of the drug was incorrect and that it had high instances of a certain side-effects. The final decision was down to Alan, and he turned down the £1million funding opportunity because although this collaboration would benefit the pharmaceutical company, the school (financially) and himself regarding 'Brownie points' or 'Kudos' for income generation, there was a chance that it may not benefit patients. Alan recognises the need to have a team around him, a team whose opinions he trusts:

> And that's why I have my team around me. I think the secret to this is that, it must never be one individual's decision.

Although Alan takes the opinion of his team on board, the final decision rests with him. We interpret this academic entrepreneur to have a unique method to facilitate sound decision making. He uses foresight to make his difficult decisions and practises what he calls 'The John Humphrys Test'. He imagines the scenario of receiving a call from an aggressive radio interviewer, to interview him regarding a project he has taken on, and he contemplates:

> Would I be comfortable to appear there and defend it. And I think, "yeah, I would, I'll take on Humphrys on that" then I'll do it. And if I think "oh, what if he asks me about that?" then I won't do it.

This is an interesting approach where the professor imagines his worst case scenario to benchmark his decision. He explains the importance of having a moral compass and explains that without one, you shouldn't be working for the university, and you may as well be working for a private corporation. The importance of individual responsibility versus organisational responsibility is also mentioned. Alan believes that being ethical and responsible is down to individual responsibility. He trusts that whether it's a commercial organisation, a university or the NHS, an organisation only functions well if the individuals within it exercise their sense of responsibility or morals. Responsibility is not just about protecting patients but is also about reputation preservation, and due to the nature of his work as an academic, he feels that; "I have a reputation, and it takes years, and years, and years to build reputation and seconds to lose it". This professor explains that funding innovation from irresponsible sources may cause reputational damage long term:

> One poor decision. I look with interest, I won't name the individual, but I've seen an individual that has worked with one student of mine in the past who would take money from anybody, I think, to do anything. And probably is making stacks of money but I'm thinking how sustainable that is in the long term, where are you going to be in 20 years' time? Might be phenomenally successful, multi-millionaire; who knows.

Here he contemplates the idea of taking research funding from companies and concludes that from a career perspective it is not worth receiving funding from just about anybody. Although Alan has been working with private companies to provide ethical, unbiased research collaborations, not all academics conduct this kind of research.

13.5   UK POLICY AND RI FACILITATION

With the introduction of Academic Health Science Networks (which facilitate collaborations between universities, hospitals and industry) and UKRI in England we are seeing a continued drive from the government to encourage academics to conduct collaborative projects with private companies, in an effort to boost the economy. Furthermore, the exit of Britain from the EU is likely to result in a reduction in available UK research funding. This will put pressure on researchers, and leave academics with no choice but to collaborate with industry. If academic–corporate collaborations increase, academics without corporate collaboration experience are likely to face unique hurdles. The increase in academic–corporate collaborations are likely to boost the usefulness of the RI practices. There is currently no RI framework which specifically aims to support the academic–corporate collaboration. The current practice and policy context within the UK may augment the usefulness of RI and facilitate the creation of a framework or checklist that may be useful to aid successful, stakeholder-centric collaborations.

There are many examples of how RI practices have been utilised effectively by Alan. These practices do not appear to have hindered digital technology success, to the contrary, they appear to have facilitated the creation of digital technology, by stakeholders and for stakeholders, in a responsible way. However, despite the successes of collaborative relationships, they have also caused tensions as parties disagree on the needs of the company and the needs of the patient, creating a level of separation between the collaborating parties. In this case the tension has come about through end-user and stakeholder engagement and discussion. Therefore, this separation could be expressed as the outcome of successful RI.

We have seen the dark side of RI in this case study also, through IP theft, despite the presence of an NDA. This demonstrates a barrier to reflexivity in the collaborative academic–corporate relationship. This case study identifies that theft was possible through corporate knowledge that universities have insufficient funds to fight IP law suits. One way to solve this problem without jeopardising collaborative RI would be the creation of a centralised UK fund which can be accessed by academic institutions to fight cases of IP theft. This in itself may act as a deterrent to the theft of academic IP and facilitate the longevity of corporate–academic collaborations.

13.6   CONCLUSION

We conclude that this narrative portrays a professor conducting responsible research, either knowingly or unknowingly. There seem to be many factors which may have affected Alan's success; his soft skills such as trust building, negotiation and convincing, his previous experience within the pharmaceutical industry and his drive to generate income to maintain his position, that is, the psychological contract. The narrative indicates that Alan's responsible approach may well have facilitated his success and the commercialisation of his products or ideas. The involvement of stakeholders, the responsiveness to their suggestions and his willingness to switch his product from one that was initially anticipated to another based on stakeholder response, have all played an important role in the success of his digital solutions. We set out in the introduction that stakeholder involvement is important in digital innovation. We interpret Alan's RI-like approach has had a positive impact on the uptake of his products. Through proper stakeholder involvement, his products have been suitably built for their context, not just technically but practically also, which has driven their success. It is recommended that innovators adopt the RI approach, to generate technologies which are desired by society, are shaped by stakeholders and are therefore useful and profitable. It appears from this case study that anticipation is the least important factor in innovation for this professor and stakeholder involvement is the most important. In this way this narrative supports Clarysse, Tartari and Salter (2011) which shows that individual-level attributes and experience are the most important predictors of academic entrepreneurship success with TTOs playing a lesser role and lacking in terms of commercial support for inventive academics, which demonstrates a lack of stakeholder understanding on the part of TTOs, and therefore a lack of RI practice.

The case for employment by results can be a successful approach for a university. This coupled with the risk-taking entrepreneurial nature of the academic helped to make this relationship possible and prosperous, with the psychological contract playing an important role as continued success was required to gain a professorial chair. One of the ways he demonstrated success was through responsible innovation. His products were honed by stakeholders and proved useful and sought-after. The psychological contract between the university and the professor resulted in a unique relationship of trust built on past success with no requirement to maintain the professor's contract. There appears to be a link between psychological contracts and the academic entrepreneurs' attributes which has culminated in digitalisation success. Working with RRI in mind, university academics can create technology which is responsible and successful. In this case, RRI has shown to be financially favourable for the university and the academic. However, true success in a university appears to be best delivered through a trusting relationship between the employer and the employee. If a psychological contract exists between a university which promotes RI practice, it is argued that this better facilitates the cascade of RI practice to professorial staff.

13.6.1 Implications for Theory and Practice

The concept of a professor existing on a temporary contract that is renewed only when he generates a certain income is a new phenomenon. It is expected that the fear of losing his position may have helped to drive his initial success; bringing in £32 million over his career. It would be interesting to understand if this model has ever been represented at other academic institutions and it thus affects the success or commercialisation of digital healthcare innovation.

This narrative corroborates with Clarysse, Tartari and Salter (2011) that technology transfer offices do not appear to provide the support that they are expected to, and therefore there needs to be a fundamental shift in the way that these hubs operate. Perhaps these hubs are recruiting staff with the wrong skill sets and motivations or maybe the right financial incentives are not provided to TTO staff to encourage them to invest motivated time into new spin-outs. In the meantime, academics looking to innovate within the university context may find it useful to consider this professor's experience. An executive education programme could help bridge the gap between academia and industry, providing training for academics in the areas of negotiation, fundamental contract law and IP. This is likely to support academics who wish to be involved in private company collaborations but feel unprepared or scared to do so. It is expected that an executive education programme would not only prepare academics as described but may help to normalise the culture of conducting industry–academic research collaborations and therefore encourage more academics to generate research income from private industry.

This narrative also supports the notion that RI must be embedded in the overall university strategy, in order for it to work as a guideline for how universities shall contribute to solve societal problems. Strategy is one of the entrepreneurial architect dimensions that lay out how universities can reach their third mission to stimulate and sustain economic development (Foss and Gibson, 2015a). This case study illustrates that there is no overarching university strategy that backs up the professor's responsible innovation mission. We suggest that policy for responsible research and innovation should be integrated in universities' institutional planning documents, incentives and policy. This narrative also supports the role of culture as a significant entrepreneurial architect dimension (Foss and Gibson, 2015b). Universities need to work on a positive academic culture, where including stakeholders and being receptive to their needs becomes an integrated part of commercialising science and technology.

REFERENCES

Clarysse, B., Tartari, V. & Salter, A. (2011). The impact of entrepreneurial capacity, experience and organizational support on academic entrepreneurship, *Research Policy*, 40, 1084–1093.

Denscombe, M. (1998). *The Good Research Guide For Small Scale Social Research Projects*. Milton Keynes: Open University Press.

Flyvbjerg, B. (2006). Five misunderstandings about case-study research. *Qualitative Inquiry*, 12(2), 219–245.

Foss, L. & Gibson, D.V. (2015a). The entrepreneurial university: context and institutional change. In Foss, L. and Gibson, D. V. (eds), *The Entrepreneurial University. Context and Institutional Change*. London: Routledge, pp. 28–44.

Foss, L. & Gibson, D.V. (2015b). The entrepreneurial university: case analysis and implications. In Foss, L. and Gibson, D. V. (eds), *The Entrepreneurial University. Context and Institutional Change*. London: Routledge, pp. 249–276.

Hendy, J., Reeves, B. C., Fulop, N., Hutchings, A. & Masseria, C. (2005). Challenges to implementing the national programme for information technology (NPfIT): a qualitative study. *British Medical Journal, 331*(7512), 331–336.

Klofsten, M. & Jones-Evans, D. (2000). Comparing academic entrepreneurship in Europe: the case of Sweden and Ireland. *Small Business Economics*, 14(4), 299–309.

Krabel, S. & Mueller, P. (2009). What drives scientists to start their own company? An empirical investigation of Max Planck Society scientists, *Research Policy*, 38, 947–956.

Lambert, R. (2003). Lambert review of business–university collaboration: final report (SSRN Scholarly Paper No. ID 1509981). Social Science Research Network, Rochester, NY.

Nelles, J. & Vorley, T. (2010). Constructing an entrepreneurial architecture: an emergent framework for studying the contemporary university beyond the entrepreneurial turn. *Innovation of Higher Education*, 35, 161–176.

Rasmussen, E. & Borch, O. J. (2010). University capabilities in facilitating entrepreneurship: a longitudinal study of spin-off ventures at mid-range universities. *Research Policy*, 39, 602–612.

Robinson, S. L. & Rousseau, D. M. (1994). Violating the psychological contract: not the exception but the norm. *Journal of Organizational Behavior*, 15(3), 245–259.

Roessner, D., Bond, J., Okubo, S. & Planting, M. (2013). The economic impact of licensed commercialized inventions originating in university research. *Research Policy*, 42, 23–34.

Rousseau, D. M. (1989). Psychological and implied contracts in organizations. *Employee Responsibilities and Rights Journal*, 2(2), 121–139.

Siegel, D. S. & Wright, M. (2015). Academic entrepreneurship: time for a rethink? *British Journal of Management*, 26(4), 582–595.

Stilgoe, J., Owen, R. & Macnaghten, P. (2013). Developing a framework for responsible innovation. *Research Policy*, 42(9), 1568–1580.

UK Research and Innovation. (n.d. (a)). Industrial Strategy Challenge Fund. Retrieved 13 July 2018, from https://www.ukri.org/innovation/industrial-strategy-challenge-fund/.

UK Research and Innovation. (n.d. (b)). Retrieved 23 August 2018, from https://www.ukri.org/.

APPENDIX 13.1: INTERVIEW GUIDE

Baseline Question

1. Can you tell us about the product or service you offer in your department, company or organisation? Years in business, employee numbers. How was it formed? Who's idea was it?

Early Market Success

1. How satisfied are you with the sales of your product/service?
2. How satisfied are you with income you have received from sales of your newly developed product/service, in relation to time, resources and efforts you used to create it?
3. Do you feel that your product/service is a success, or you are satisfied with progress? If we ask you to rate it from 1 to 10, with 1 being a failure and 10 being super success, what rate would you give?
4. Are you satisfied with the increases in sales? With profits?

Inclusivity (RI) – How Do You Include Your Stakeholders?

1. How did you understand what was important to the user – what is it that makes you understand what is important?
2. Was there anyone else who made you change your view? How did this happen?
3. Who did you first include when you started thinking about the idea? Why this person?
4. Do you consciously decide who you include in product development? What are your criteria for inclusion?
5. How important is the user for you in the development of the product or service?

Reflexivity (RI) – How Do You Reflect on Your Firm's Success? And How Do You Change Your Approach as a Result of Reflecting?

1. What was your motivation for this product/service when you started and what is your driver now?
2. What was the problem you were trying to solve?
3. Was there anything you wasted a lot of time on?
4. What was your original plan for the commercialisation? How did your reflections change the way the product commercialised?
5. To what extent is there room for reflection in the company's daily operations?

Responsiveness to Stakeholders, and External and Internal Factors, Responsive to Negative Consequences (RI)

1. Whom do you consider as your stakeholders, can you describe them?
2. Is there room for change based on input from customers, users, or other stakeholders? (How have the stakeholders been involved?)
3. Has the idea changed in relation to the initial plan based on stakeholder input? (Describe)
4. At what stage (beginning, later) was it important to include users? And why at this stage?
5. Do you receive praise from stakeholders for the way you interact with them?
6. Do stakeholders agree with what your product is trying to do?

Anticipation (RI)

1. What is the consumer's expectation of the product/solution?
2. Will the solution/product help solve a societal problem? (Describe)
3. Can you describe the ideal social impact of the product?
4. Did you develop the product/solution you intended to create? If you didn't, what did you envision to be the ideal solution? What were your assumptions around that product/solution?
5. Who did you envision your customer/user to be initially? Did it differ? How?
6. What challenges/obstacles did you experience – economic/social. Was this different from what you initially thought?
7. Was the problem that you aimed to solve different than what you expected?
8. Can you tell us something that really surprised you about the product/ service?

Responsibility (RI)

1. How important is it for you to follow the community's formal and informal rules and laws?
2. How important is it for you to respect the thoughts and ideas of others?
3. How important is it for you to take responsibility for the community's concerns?

Entrepreneur's Perspective on RI

1. What is your perception of Responsible Innovation?
2. We just spoke about different factors that related to responsible innovation, i.e., being ethical, involving your stakeholders, being reflective with a socially responsible aspect to your work.
3. Generally do you think that responsible innovation affects the success or commercialisation of a firm or product? In what ways and does it affect you? Or might it in the future?
4. Are there ways that you think responsible innovation can facilitate success, or are there any obstacles to conducting responsible innovation?

Other Barriers to Commercialisation

1. What other barriers have you experienced starting a business from a university department?
2. Does being a department unit versus a spin out company affect how well the company does?
3. Institutional barriers, structure of the 'company', – how do these affect success?

14. Responsible innovation within the healthcare sector: digital therapeutics and WellStart Health

Jill Kickul, Mark Griffiths, and Marissa Titus

14.1 INTRODUCTION

It is not a revelation that the healthcare system within the United States is in dire need of disruption. Relative to its size, the United States spends a disproportionate amount on healthcare with little to show for it. As of 2016, the United States leads in per capita healthcare spending at $10,348. This is 31 per cent higher than the second leading country, Switzerland, which spends $7,919 (Sawyer and Cox, 2018). Nevertheless, the United States is only ranked 42nd in global life expectancy (Reynolds and Avendano, 2018). How is it that a country that spends so much can rank so poorly? To create a healthcare system that is both more effective and affordable, *disruptive innovation is imperative*.

A growing sector of health innovation involves digital innovations that offer the ability to decrease costs, increase access, improve effectiveness, and closely monitor individual outcomes. An emerging segment within this space is digital therapeutics. Digital therapeutics are software-based interventions that help treat diseases by positively changing individuals' behaviours and closely tracking outcomes. These new developments in the delivering of healthcare are an outgrowth of an earlier development: telemedicine. As Pisacane (1995) points out: "Telemedicine, literally medicine at a distance, is the delivery of health care over long distance using medical knowledge combined with communications and computer technology."

The first aim of this chapter is to examine the innovation environment and assess how it may benefit from the principles of responsible research and innovation (RRI). The RRI approach develops innovations with the impact on the stakeholders as the central consideration. The second aim of this chapter is to analyse one company operating within this digital therapeutic arena, WellStart Health. Founded in 2015 by Olivia Kelly, MPH and Bojana

221

Jankovic Weatherly, MD, MSc, WellStart Health's mission is to leverage its physician-led digital therapeutic platform to reverse chronic disease.

14.2   THE HEALTHCARE INDUSTRY WITHIN THE UNITED STATES

The healthcare system in the United States is in a major crisis. The sector faces unprecedented challenges that are magnified by the aging population, increased budget constraints, and desire to provide care to all. Further, the patients themselves are changing. They are becoming increasingly well-informed about their health status and have come to expect an open and interactive relationship with their medical providers. In general, there is an overwhelming increase in the desire to play an active role in managing their individual health.

Currently, the healthcare system is both labour- and resource-intensive. There are numerous stakeholders ranging from the patient, to the provider, and to the payer (often insurance companies or the government). As a consequence, there are numerous competing interests, which must be considered. How can innovations decrease the cost of healthcare while maintaining or ideally improving quality? Solutions must strive to address both of these competing interests.

Innovations in information and communication technologies (ICT) have the potential to contribute to the development of solutions to many of these challenges. Specifically, telemedicine is promising as it can improve access, decrease costs, improve quality of care, and directly affect individual patients. Simply defined, telemedicine is the use of information and communication technologies to provide clinical healthcare at a distance. This aids in eliminating the barrier of distance and improving the access to service.

Dogaru, Stanescu and Luminita (2014) report that: "Experts predicted as early as 10 years ago that starting with 2010 at least 15 per cent of health care services worldwide would be provided from a distance using telemedicine" (Sinha, 2000).

The desire to provide healthcare to those with limited access to medical assistance is not really a new innovation or idea. On 15 May 1928, the Reverend John Flynn received a large bequest which enabled him to open the Australian Inland Mission Aerial Medical Service (later renamed the Royal Flying Doctor Service [RFDS]). The first Flying Doctor took flight on 17 May, 1928. At the time, the Flying Doctor Service lacked the communication technology to deliver services efficiently. This barrier was overcome with the invention of a pedal-operated generator to power a radio receiver. By 1929, people living in isolation were able to call on the Flying Doctor to assist them in an emergency. Inventor Alfred Traeger devised a generator using bicycle pedals allowing the operator to pedal to provide power and leave their hands free to send Morse code. In all, Traeger built 3,000 pedal sets and later designed and built a simple radiotelephony set allowing outpost stations to communicate by voice instead of Morse code. The Royal Flying Doctor Service remains in service to this day and covers more than seven million square kilometres, a territory larger than Western Europe or almost 73 per cent of the size of the United States.

Despite its potential, innovation within the healthcare space has proven slow to implementation and commercialization. A key barrier to innovation is the fear of change as it threatens the existing business models (Christensen, Bohmer and Kenagy, 2000). As Pisacane (1995) pointed out, costs, reimbursement policies, liability concerns, insufficient standards and technological limitation constrain the growth, development and utilization of modern methods of delivering medical services. That is, powerful institutional forces exist opposing these changes implemented because they threaten existing business models. Note that this is unlike the RFDS, where no competing model existed. While this is understandable from the provider's perspective, it is not in the best interest of the patient.

14.3   RESPONSIBLE INNOVATION

As previously described, innovation within the health sector is complex due to the multifaceted nature of its stakeholders. Consequently, the sector would benefit from an approach to innovation that is designed to take into account each of these actors. Responsible Research and Innovation (RRI) is one such framework. The basis for RRI is that it approaches innovation by assessing and anticipating the potential implication as well as societal expectations with regard to innovation and research. It aims to design innovation in a method that is both inclusive and sustainable (European Commission, Horizon 2020, 2018). Von Schomberg (2013) defines RRI as:

> Responsible Research and Innovation is a transparent, interactive process by which societal actors and innovators become mutually responsive to each other with a view to the (ethical) acceptability, sustainability and societal desirability of the innovation process and its marketable products.

A key benefit to responsible innovation that increases its appeal within the healthcare sector is that it closely considers the perspective of the user. Therefore, this increases the marketability of the innovation in addition to the likelihood of successful adoption and institutional or systemic change. Innovations that offer a competitive advantage alone are not sufficient to gain commercialization as they must be embraced by the individuals and become embedded in society. RRI involves all societal actors throughout the innovation process to align the outcomes with the central values and expectations of society.

The responsible research and innovation process has been broken down into four crucial areas: anticipation, inclusivity, reflectivity, and responsiveness (Owen, Bessant and Heintz, 2013). Anticipation involves determining whether the potential implications of an innovation have been thoroughly assessed. Inclusivity examines who the end user was viewed to be. How was this person included throughout the innovation process? Reflexivity addresses actively seeking feedback regarding the innovation. Finally, responsiveness assesses whether or not the innovation was sufficiently flexible and open to change throughout the process. Was there a willingness to adapt to user feedback?

In the healthcare field, the patient is at the centre of the system. While consideration of additional stakeholders including the provider and payer is beneficial in developing solutions, the patient should remain of primary concern. This leaves the question of how can healthcare innovations be designed to serve the needs of the patient?

14.4   DIGITAL THERAPEUTICS

The need for the healthcare system to cut costs while improving outcomes has led to an emerging subset within digital health: digital therapeutics. Digital therapeutics allow for the treatment of medical or psychological conditions via the use of digital technology. The aspiration is that digital therapeutics will be able to improve one's health as much as a drug can, but without the side effects and high cost. The 'active ingredient' for digital therapeutics is the software behind them. Essentially digital therapeutics are 'software as a drug'. They have the potential to either augment or replace the current medication regimen of patients.

Digital therapeutics rely on motivating lifestyle and behavioural changes. The resulting push to help users change their lifestyle behaviour is both discrete and non-interventional. Research has shown that 40 per cent of premature deaths are directly related to lifestyle factors (McGinnis, Williams-Russo and Knickman, 2002). There are a wide range of conditions and disease that these therapeutics can be utilized for, including type II diabetes, atherosclerosis, cardiovascular disease, obesity, respiratory diseases, hypertension, hypercholesterolemia, smoking cessation, and more. The key challenge with these conditions is that these are not disease states that can be quickly medicated or inoculated away like we have come to expect with modern medicine.

The market for digital therapeutics within the United States is estimated to reach \$9.4 billion by 2025 (Grand View Research, 2017). This is a steep increase from the reported market of \$1.7 billion in 2016. This 21 per cent compound annual growth rate is fueled by an increase in the incidence of chronic disease, the need to decrease healthcare spending, and the shifting emphasis towards preventative health. Further, there is also a significant increase in the number of venture capitalists who are investing in this market.

Perhaps most interesting about digital therapeutics, is the potential to not only disrupt healthcare delivery, but also the pharmaceutical market. With healthier patients, there will be a decreased time and resource burden on the healthcare system. In addition, digital therapeutics have been found to enable patients to decrease or go off of their current medications so the cost burden associated with high medication costs will be alleviated.

14.5 BENEFITS OF DIGITAL THERAPEUTICS

The potential benefits of digital therapeutics are vast. Aside from the more evident improvement of health, they offer potential benefits to a wide range of stakeholders, generating a highly positive return for society as a whole. These stakeholders include patients, providers, and payers. Analysis of each stakeholder reveals strong benefits to each:

> *Patients:* For the patients, digital therapeutics are able to provide continuous, real-time feedback. Digital therapeutics are equivalent to having a 24/7 health coach available in their pocket. This support system increases the chance of positively adopting a healthier lifestyle. Programme adherence is easily tracked, reminders are quickly accessible, and results are closely tracked.
>
> *Providers:* Digital therapeutics offer the providers 24/7 access to their patients. They are able to continuously monitor their health and respond as needed. Providers will be able to utilize the data from digital therapeutics to see where patients are falling short. Consequently, they will be able to provide specific and actionable recommendations to directly impact the area of weakness. Providers will also be able to gain a more holistic view of behaviour change. While every patient is different, they will become more knowledgeable regarding the most difficult areas of change and what they can be doing to help.
>
> *Payers:* Healthcare within the United States has numerous potential payers. For uninsured patients, the burden falls on the patient themselves. Others may be covered by private insurance, their employers, or government agencies. Each of these players could potentially save a significant amount of money on healthcare and pharmaceutical expense. The United States government has demonstrated buy-in with this idea. In 2016, Medicare agreed to reimburse the cost of participation in Omada Health's diabetes reversal programme. If insurance companies continue to support these interventions, the space is poised for rapid growth. It is likely that they will become increasingly integrated into our current standard of care.

Another key benefit of digital therapeutics is that unlike a pharmaceutical drug, they do not require the United States Food and Drug Administration approval since they are considered to be low-risk behaviour changes. This saves billions in clinical trial expenses as well and time to market.

Finally, an enormous strength of digital therapeutics is their potential to generate highly impactful data. Access to this data can demonstrate efficacy and allow for continuous iteration of the behaviour change model to increase its effectiveness and marketability.

14.6   RISKS OF DIGITAL THERAPEUTICS

Despite the numerous benefits of digital therapeutics, it is important to mention the existence of several risks and obstacles that must be overcome in order for the industry to truly take off.

Currently, there are limited studies demonstrating the efficacy of digital therapeutics. While not requiring FDA approval helps the innovation get off the ground significantly faster compared to a new pharmaceutical, it also means that there is a lack of robust data to demonstrate the safety and efficacy of the intervention. As a result, it is left up to the companies to perform the studies themselves. Partnering with external organizations to complete the trials decreases the potential bias of internally produced data. Luckily, the digital nature of these interventions makes data collection relatively seamless to capture. Proving efficacy is essential if digital therapeutics will be covered by payers including the patient, employers, private insurance companies, and the government.

A second concern around digital therapeutics is the compliance of patients. Proper compliance with pharmaceutical therapeutics is challenging enough, but digital therapeutics are asking patients to completely transform their life-styles. Pharmaceuticals only ask the patient to remember to take a pill at the same time every day, and even that can be challenging. For digital therapeutics, completely changing ingrained human behaviours is extremely difficult to do. As such, it will likely be impossible for these interventions to work for everyone and it will require varying timeframes to occur depending on the individual's personal motivations and self-discipline.

Finally, acceptance and adoption by physicians is critical to scale this technology. Should the experts, physicians, not recognize the power of the technology; it is facing an uphill battle. Physicians could potentially begin 'prescribing' digital therapeutics the same way that they do pharmaceuticals, spurring the market. However, in order for this to transpire, it will once again be imperative that clinical evidence is available and communicated. Doctors trust clinical trials. Positive outcomes of the trials will motivate doctors to pre-scribe digital therapeutics as an intervention. Nonetheless, it seems that digital therapeutics are significantly lower risk than drugs, therefore the amount of upfront clinical evidence will not need to be nearly as high. The data generated after implementation will continue to demonstrate whether or not digital therapeutics is effective and safe.

Furthermore, physician adoption requires that they are willing to modify their current standard of care. Should digital therapeutics become the norm, it will be imperative that physicians have access to the data that is being collected. However, this requires that they have the time to assess the findings and respond as needed to the information.

In conclusion, the key challenge surrounding digital therapeutics is that they must be able to demonstrate efficacy. In order to do so, it is critical that patients are compliant with the programmes. Should these two factors fall into place, the potential for digital therapeutics to gain widespread adoption and to revolutionize the health and pharmaceutical industry is high.

14.7   WELLSTART HEALTH INTRODUCTION

Olivia Kelly was studying for her Master's in Public Health in Health Policy and Management at the University of California Los Angeles, Fielding School of Public Health when she first began conceptualizing the idea for what would become the digital therapeutic company WellStart Health. She was able to recognize that the cause of many disease states were directly related to people having an inappropriate relationship with their environment. She wanted to help people decrease their risk or even reverse their chronic diseases through intensive changes to their diet, lifestyle, and their environment. She decided to develop her model to focus on five pillars to creating optimal health: whole food plant-based nutrition, physical activity, mindfulness, social support and resilience.

Initially, Olivia envisioned WellStart Health as a bricks-and-mortar clinic where patients would come directly to participate in the clinician-led programme. She thought that the direct interaction and support system would be essential in aiding the behaviour changes. However, after sitting down to work though the concept with a strategy professor at UCLA, she quickly realized that it would not be a financially feasible model. As a result, Olivia pivoted towards a virtual model in order to overcome the financial constraints. WellStart Health officially launched its pilot programme in Fall of 2015.

14.8   WELLSTART HEALTH PROGRAMME

Olivia Kelly developed WellStart Health with a customer-focused approach. During her initial pilot she received a great deal of feedback, and updated her model accordingly. WellStart offers a B2B focused virtual health 'retreat' that allows its users to take control of their health, decrease their risk factors for chronic conditions, and even aids many users in reversing chronic conditions.

It was found that 84 per cent of users decrease their blood pressure, 35 per cent reduce or eliminate medications, and 80 per cent reverse at least one chronic condition. Across all participants, they lose an average of 6.9 pounds during the 12-week duration of the programme.

WellStart Health's programme is led by a team of clinicians that include health coaches, dieticians, and physicians. The programme focuses on five core areas: optimal diet, mindfulness, movement, resilience, and social support. A heavy focus area is switching to a diet that is more plant based. Throughout the programme, users are supported with knowledge and practical skills necessary to take control of their health. They receive access to practical tools including recipes, shopping lists, and meal plans which are delivered via the WellStart platform. Throughout the programme, participants are continuously educated through bite-sized lessons on evidence-based benefits of whole-food, plant-centred nutrition, regular exercise, and mindfulness. These lessons are delivered through articles and videos. Participants also participate in daily text exchanges with highly-trained health coaches designed to encourage goal setting as well as tracking through the platform. Additionally, there are three virtual dietitian and physician visits conducted throughout the programme which serve to offer an unparalleled level of clinical supervision. Finally, group discussions facilitated by health coaches are an integral component of the programme. Overall, WellStart Health strives to empower their participants to access the resources, knowledge, and confidence to live a healthier lifestyle, free of disease and disability.

WellStart Health measure biometrics at the start and end of the programme to demonstrate the programme's efficacy for each individual user. These biometrics include body weight, body mass index, blood pressure, cholesterol, triglycerides, and haemoglobin A1c.

Following the completion of the programme, WellStart Health offers an optional 40-week support service to ensure that the behaviour changes stick. This support service is focused on providing check-ins and group discussions.

WellStart Health leverages its data to assess and predict the methods that work best for different participants. They are able to adapt their model to incorporate these findings and continuously improve.

For Olivia, technology has enabled her to be closer to the end-user. Throughout the day she is able to have multiple touchpoints with the users to continuously support and motivate them on their journey. Still, Olivia states that they are actively trying to identify the optimal number of contact touch points throughout the day.

While WellStart Health does offer business to consumer purchasing options, the company has identified larger, self-insured employers as a key market opportunity. As healthcare expenses continue to rise, employers are looking for ways to decrease their healthcare costs associated with managing chronic conditions amongst their workforce. By improving their employees' lifestyles, research has shown that they are able to decrease their medications and potentially eliminate them.

To date, the physician feedback received regarding the programme's efficacy is highly positive. One Internal Medicine physician at Cedars-Sinai, Dr Chris Fitzgerald, stated that:

> WellStart Health will seamlessly integrate into patient's treatment plans and give patients what they want and need. The pilot shows that WellStart not only improves patient biometrics, but most importantly WellStart finally gets at the core issue that medicine is struggling to address: improve the quality of patients' lives. Primary care doctors need help, and WellStart Health provides an easy solution.

14.9 WELLSTART HEALTH PARTNERSHIP

In April 2018 it was announced that WellStart Health would merge with the Big Change Program. The Big Change Program is another 12-week lifestyle intervention programme designed to incorporate healthy habits in a powerful and sustainable fashion. This programme is focused on the pillars of science, simplicity, and sustainability. The merger was designed to expand WellStart Health's offerings for health coaching and corporate health improvement. This merger will add additional experience to their team of physicians, public health professionals, and behaviour change experts. The incorporation of the Big Change Program into WellStart Health's platform will enable the company to reach a significantly larger number of people.

Founder Olivia Kelly states:

> We founded WellStart to shake up the healthcare industry by delivering pre-primary care with the help of technology. That means offering the right guidance and support to tackle the root causes of chronic disease. With the addition of the Big Change Program's innovative behaviour change methodology and expert team, we're now able to scale our efforts more rapidly, and bring these benefits to many more people.

14.10 WELLSTART HEALTH CHALLENGES

Kelly recognizes that WellStart Health is still in its nascent stages. Each day they must continue to prove themselves and refine their model in order to grow sustainability. While WellStart Health does have available research-based evidence of the programme's effectiveness, they are looking to continue to build this data. WellStart Health was set to partner with a research university in order to conduct a clinical study to demonstrate the effectiveness of the programme. However, the research university backed out at the last minute. Kelly believes that they were not equipped for the high requirements of the study per the protocols, but she is continuing to look for another organization to conduct the trials.

Another key challenge is competing against firms who have greater funding. WellStart Health has raised an initial seed round of $300,000. According to Crunchbase, competitors such as Omada Health have raised $126.5 million, Proteus Digital Health has raised $422.3 million, and Virta Health has raised $82 million.

Finally, Kelly stated that she still does not know how the consumer adoption will be when it is provided by the employer. This month they secured their first employer contract, but they are not yet far enough into the programme to assess its efficacy. The decreased 'ownership' of being provided a programme rather than seeking it out and paying for it themselves will present an interesting factor.

14.11 CONCLUSION

While healthcare has a culture of moving slowly and cautiously, the enormous cost and resource pressures that they are currently facing presents the opportunity for the emergence of innovations offering very rapid change. Digital therapeutics have the potential to be one of these transformative innovations. They are positioned to be able to increase the quality of care while simultaneously decreasing the costs associated with it.

Digital therapeutics is currently in its early stages. The technology is not the barrier to adoption. This exists and is easily utilized. The two central challenges are proof of efficacy and consumer habits. The success rates and communication of this information to the vast number of stakeholders involved will spur the spread of this innovation. However, it fully rides on the ability of the consumer to change their habits. Not only must they be willing to consider the potential of digital therapeutics as "medicine", but also to adopt the necessary lifestyle changes necessary to follow the requirements set forth by the respective programmes.

REFERENCES

Christensen, C. M., Bohmer, R. & Kenagy, J. (2000). Will disruptive innovations cure health care?. *Harvard Business Review*, 78(5), 102–112.

Dogaru, G., Stanescu, I. & Luminita, P. (2014). Medical telerehabilitation: an effective method for the long-distance functional evaluation of patients. *Balneo Research Journal*, 5(1), 21–24.

European Commission, Horizon 2020 (2018). Responsible research and innovation – Horizon 2020 – European Commission, 12 June. Retrieved

6 December 2018 from https://ec.europa.eu/programmes/horizon2020/en/h2020-section/responsible-research-innovation.

Grand View Research (2017). Digital Therapeutics Market Worth $9.4 Billion By 2025 | CAGR: 21.0%. Retrieved July 2017 from https://www.grandviewresearch.com/press-release/global-digital-therapeutics-market.

McGinnis, J. M., Williams-Russo, P. & Knickman, J. R. (2002). The case for more active policy attention to health promotion, *Health Affairs*, 21(2), 78–93.

Owen, R., Bessant, J. & Heintz, M. (eds) (2013). *Responsible Innovation: Managing the Responsible Emergence of Science and Innovation in Society*. Chichester: John Wiley & Sons.

Pisacane, V. L. (1995). Telemedicine: health care at a distance. *Johns Hopkins APL Technical Digest*, 16(4), 373–376.

Reynolds, M. & Avendano, M. (2018). Social policy expenditures and life expectancy in high-income countries. *American Journal of Preventive Medicine*, 54(1), 72–79.

Sawyer, B. & Cox, C. (2018). How does health spending in the US compare to other countries?. Peterson-Kaiser Health System Tracker. Menlo Park, CA: Kaiser Family Foundation.

Sinha, A. (2000). An overview of telemedicine: the virtual gaze of health care in the next century. *Medical Anthropology Quarterly*, 14(3), 291–309.

von Schomberg, R. (2013). A vision of responsible research and innovation. In Owen, R., Bessant, J. and Heintz, M. (eds), *Responsible Innovation: Managing the Responsible Emergence of Science and Innovation in Society*. Chichester: John Wiley & Sons, pp. 51–74.

15. The future of responsible innovation

John Bessant, Tatiana Iakovleva and Elin M. Oftedal

15.1 INTRODUCTION

We began with a crisis and a possible resolution in the area of healthcare. Healthcare has improved beyond recognition over the past fifty years but so too have the challenges. An ageing population, concerns over lifestyle diseases like diabetes, increasing pressures on mental health, and so on all add to the burden of systems which struggle to find the resources to deliver on these increased expectations. Payment and reimbursement models of various types face the same challenge. Demand outstrips our ability and willingness to pay.

Into this crisis comes a powerful actor – digital healthcare technology. Across a broad frontier this field is opening up significant lines of attack – in education, in prevention, in treatment and in the overall management of healthcare systems. In hospitals, labour accounts for more than half of most health operating costs, and up to 90 per cent of variable costs (Ronksley et al., 2015). It is important to remember that some of these require administration of patients and clinical personnel on a scale which dwarfs even the largest organizations in the world. It is estimated, for example, that the UK NHS is the world's second largest employer with over 1.5 million staff – only beaten by the People's Liberation Army in China.

Digital healthcare holds out significant promise, offering to square the circle and allow better quality at lower cost. More advanced scheduling apps, for example, can allow nurses and other managerial staff to circulate schedules on staff mobile devices and update them in real time. Each day brings news of another technological advance and the implications for a rosy future are clear. Whether it is a better informed and empowered patient base, robot staff and cares, automated processes or personalized medical care – there is about digital healthcare the same sense of optimism as once fueled dreams of the automated factory (Blaya, Fraser and Holt, 2010).

But into this space it is important to inject a note of caution. Technological change is not neutral – it is the result of actions and decisions taken by many actors reflecting different degrees of stakeholder interest. The existing structures are controlled by the strongest and most powerful stakeholders. The patient is, as stated in Chapter 3, without strong powerbases. Since dominant designs and technological trajectories emerge it is hard to slow down. As with any major innovation which has a bearing on our well-being and quality of life, it is important to reflect on the ways in which it is being designed and implemented and to ask some cautionary questions at an early stage, before the trajectory becomes too entrenched.

This approach – 'responsible innovation' is an old one with a long pedigree of research and, more important, a range of tools and frameworks with which to approach the challenge. Technology assessment, simulation, forecasting – all play a role in helping shape innovation for the greater good. In the 'mainstream' world of commercial innovation where there is a clear market, it makes sense to deeply understand the needs of that market, including its latent and unarticulated needs. Tools like empathic design, design thinking, lead user, and so on have augmented the traditional approaches of market research and provide powerful mechanisms through which the 'voice of the customer' can be heard and translated into products and services which really do meet their needs.

The same is true of healthcare with much innovation mediated through clinicians. Within this space there is scope for considerable involvement; indeed healthcare innovation has more than its share of user innovators who may well develop early prototypes (Oliveira et al., 2018). Given the size of this market it is not surprising that the 'supply' side for digital healthcare has expanded considerably with both start-ups and mainstream players offering a growing range of solutions. But arguably the problems which they work on and the understanding of user needs comes in a two-step process in which clinicians act as expert proxies for those users.

15.2 DESIGN SPACE AND RESPONSIBLE INNOVATION

In the first part of the book we set out the potential challenge – at the limit digital healthcare risks opening up a digital divide, offering significant advantages on the one hand but disempowering end-users on the other. We introduced the concept of 'design space' – essentially the arena within which the discussions and debates around the shape and direction of technological change become hammered out. Maintaining as wide a design space for as long as possible is a desirable outcome in any decision process regarding new technology, and history has some important lessons to teach us about the benefits of keeping it open and the costs of closing down too soon. Arguably 'responsible innovation' (RI) is about holding the space open for long enough to explore and enable better (= more inclusive) designs and solutions.

We pointed that innovations are embedded into the existing context of the different health systems. Innovators, whether they are start-ups, established companies or individuals, are embedded in a complex mix of government actors, insurance companies and providers, where laws, norms and knowledge form a strong institutional framework with specific selection mechanisms.

We also introduced a framework for RI based on current literature (Stilgoe, Owen and Macnaghten, 2013; Stahl et al., 2017) focusing on purpose, process and outcome. While the purpose distinguishes this type of innovation in terms of responsibility, in the way that it is an innovation borne out of a desire to solve a societal challenge, the process determines the responsible way of reaching the specific goal. Finally, the outcome has to be balanced with the purpose, unless we only have a set of goals with no implementation. This helpfully brings up to date the long-standing theoretical and practical challenges involved. In this book, we have selected cases out of purpose, so we are focusing especially on the framework originally developed by Jack Stilgoe, Richard Owen and Paul Macnaghten (2013). Their four 'tests' help focus analysis of innovation processes in terms of the degree to which they explicitly address concerns about responsibility. These are: inclusiveness, anticipation, reflexivity and responsiveness.

Inclusiveness of different stakeholders in the innovation development process can be seen as one of the crucial tasks for firms to achieve sustainable growth and successful launch of their innovations into the market. Anticipation includes the process of foreseeing side effects of the innovation or the firms' action and can be seen as a type of a risk assessment. Reflexivity advises on an ongoing assessment of the underlying foundation of the innovation or operation. Responsiveness means a capacity to change shape or direction in response to stakeholder and public values and changing circumstances.

15.3   TOWARDS THE EMPOWERED PATIENT

There is a growing concern that in this process the needs and indeed the experience of the patient as end-user is not reflected adequately. Concern is emerging that the 'voice of the patient' is not being heard and this has implications for innovation processes. At the same time the rhetoric amongst healthcare providers and their governing authorities is increasingly towards patient inclusion and empowerment.

This welcome emphasis on patient inclusion is valuable but it raises the question as to the extent to which patients wish to or are capable of being involved, especially in the design of technological innovations. As we discussed in Chapter 3, there is scope but we should take care to recognize that not all patients wish for more than to be passive recipients of healthcare. However there is a significant number – as our cases confirm – who wish to play a more active role in the development of innovations. We believe it is important to move the patient empowerment debate from considering patients as a homogenous group to viewing them as distributed across a spectrum.

The mission of this book is to understand how digital tools can give the patient an arena to actually exercise a bit more user power. We believe that RI principles can be used to empower the patient. The patient is empowered on a continuum of involvement and distinguished as three archetypes, as demonstrated in Figure 15.1.

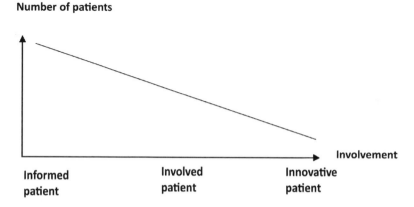

Figure 15.1   *Spectrum of patient innovation behaviour*

1. The 'informed patient', equipped to use technology based on improved understanding; not only are today's patients able to search for information with regard to their situation, they can also become active discussants of their situation with healthcare professionals.
2. The 'involved patient', playing an active role within a wider healthcare delivery system and enabled to do so by technology. Here the approaches widely used in the commercial sector are finding increasing application with users actively engaged at the 'front-end of innovation', evaluating prototypes, providing valuable feedback to help pivot designs and acting as a 'crowd-sourced' laboratory for development.
3. The 'innovating patient', providing ideas of their own based on their deep understanding of their healthcare issue. At the limit we find here the kind of patient described in Chapter 4, active 'hero' innovators, prototyping and trialing their ideas out on themselves or their nearest and dearest.

The 'inclusiveness' dimension of RI is thus of particular interest; it follows that addressing the concerns of patients and finding ways to build these insights into designs will be of paramount importance. But how to enable such participation? How to articulate needs? How to move patients along the spectrum from passive consumers to involved, active or even innovative players in the innovation game?

We continue with the discussion of four RI dimensions: inclusiveness, anticipation, reflexivity and responsiveness, as useful tools to answer the above questions.

15.3.1 Inclusiveness

Our cases confirm the view that digital health opens up significant opportunity for increasing participation of multiple stakeholders – including patients themselves. In almost all of our cases patients are recognized as an important potential knowledge source, and there is evidence that design methods are increasingly being used to ensure their input is captured at an early stage. Chapters 4–14 provide examples of such inclusion in different degrees. As we have seen, the nature of innovation processes based on agile/LSU models mean increasing use of early stage prototypes and MVPs followed by user-inspired pivoting.

But we need to look in more fine-grained fashion at the ways in which users are being included. Our spectrum is helpful here, and it will be useful to summarize our findings around the three points along that spectrum characterized by different patient archetypes.

Informed patient
At the minimum there is scope for the *informed* patient – indeed without their approval system changes like EHR will not be possible. Much of the rhetoric around healthcare today speaks of the need to bring the patient in but often points to this role rather than the richer and more active one potentially played by innovative or involved patients. But moving patients along the spectrum towards these more active inclusion positions will take tools, training, and time.

The concept of an informed patient – a patient who is knowledgeable about his or her condition and engaged in health decisions – is emerging. This can ease the job for the doctor and bring more knowledge into the health system. This is illustrated well in the MyChart example from the US (Chapter 6), where patients have better information access to their health records and communication with healthcare institutions.

Questions here are raised about how to inform and empower patients – and one of the promising directions which digital innovation can help with is in enhancing this. For example the original idea behind Healthcare TV was to empower patients – although the innovation itself came about by the actions of a hero innovator, developing a prototype on the back of the poor experience of his father, the desire to create a more open easy access way of informing patients has considerable potential.

As demonstrated in Chapter 5, patients do provide a lot of feedback to clinicians. However, reflecting on the feedback requires system changes, which are not always supported at a current stage. Patient feedback enhanced by digitalization for the sector might bring considerable value, enhancing measurable outcomes such as quality of care and availability of care.

Involved patient

Although there are some examples of 'innovating patients' the majority of our cases are about working with 'involved patients', able to be co-creators, contributing to the innovation process at both the design and diffusion ends of the journey.

A priori the case for involving patients is strong, not least because they help in identifying the questions to ask and the outcomes to assess (Sacristán et al., 2016). And one of the valuable features of digital technologies is the possibility of embedding such insights at an early stage; since the technologies are programmable and re-programmable there exists scope for co-creation during the development phase. Such approaches have increasingly been used in improving processes and services, drawing on design thinking tools to enable a better understanding of 'the patient journey' and adapting delivery systems to those insights (Pagliari et al., 2005).

However, user feedback and ideas have to be heard and transformed into outcomes. Thus, this co-creation process is dependent on three factors:

- The ability and willingness of patients to get involved and contribute their insights.
- The willingness of the innovating organization to absorb and react to this type of feedback.
- The flexibility of the technology – the degree to which it can be shaped and modified in response to patient insights.

Clearly service and process design are more amenable to such in-line modification; product design is often less flexible. But as we noted above a key feature of digital technologies is their inherent flexibility and the potential to 'delay the freeze' in terms of final design and production at scale.

To some extent there is a dynamic visible in our case studies which drives towards greater patient involvement because it leads to better design and downstream outcomes. The case of the academic entrepreneur in Chapter 13 is a good illustration; here the original approach was to a large extent driven by interactions with clinicians. But experience with users, especially feedback from asthma patients, led to revisions not only of that specific design but of the subsequent approach taken by the innovator.

One of the important aspects of working with involved patients is that they provide a helpful coalition through which to learn and develop novel technologies. The example of digital therapeutics (Chapter 14) is a good example; here the opportunity to work with something which has considerable potential as a disruptive innovation is enhanced by an active involved user base. Christensen's prescription in such situations is not to work with the mainstream but rather to find ways of exploring at the fringes – and gaining valuable experience by working with users (Christensen, Waldeck and Fogg, 2017).

So the question is less one of whether patients have the knowledge than about *how to enable* its articulation and transmission at an early enough stage in the design process. A variety of studies (Kamper, Maher and Mackay, 2009; Bessant and Maher, 2009; Bate and Robert, 2007; Bevan and Hood, 2006) suggest that design thinking and methods provide a rich toolbox and there is evidence amongst our cases of using this kind of approach. There is probably scope for more – and in the process we might well be able to move patients from the passive to the active end of the co-creation spectrum by giving them a voice.

One major challenge, especially for start-ups lies in their intersection with the wider healthcare establishment. Most of our start-ups recognize the potential value in working with users – but their experience, as illustrated in Chapter 10, is that they often come up against health systems which are closed to them and where their ability to bring in RI solutions is to some extent constrained by institutional walls.

Innovating patient

As von Hippel and colleagues have demonstrated over many years, users are a powerful source of insight at the front end of innovation (von Hippel, 2005). In particular, a combination of their high incentive to innovate and their tolerance for imperfection means that early-stage prototypes often open up radically new directions for innovation – and because of their origin also open up a wide potential downstream marketplace for 'people like us'. Diffusion is linked to compatibility and being able to meet precisely the needs of users is a powerful accelerator for adoption.

Much of this depends on the articulation of 'sticky information' – the tacit knowledge which may not be evident to external agents nor even to advanced market research techniques (Reber, 1989). Users know, by definition, what it is they want and the conditions under which it will work – the challenge is finding ways of capturing and working with that.

So user-driven innovation is a potentially rich resource – and nowhere more than in healthcare. Who knows the challenges of living with diabetes or cancer better than those who experience it? Patients with chronic diseases are often forced to find solutions for their conditions, if solutions for their problems are not available already on the market. There is growing evidence that many patients and caregivers develop solutions to cope with their health disorders (Canhão, Zejnilovic and Oliveira, 2017; Zejnilovic et al., 2016).

But evidence also shows that these innovations have low diffusion. This is, especially, a situation in the segment of rare diseases and chronic needs – niche markets, unattractive to stakeholders in the healthcare industry – where patients are pioneers regarding innovation when compared to commercial producers, which can help close gaps in the delivery of medical care. Evidence suggests that patients also adopt solutions developed by other peers (patients or caregivers), as illustrated though our case Patient Innovation in Chapter 4. They organize into communities, individually and jointly solve problems, share solutions among peers, and even undertake limited trials with the solutions they develop. However, in sharp contrast to the frequency of patient-to-patient sharing, only 6 per cent of patients reported describing their innovations to their clinicians, where reduced appointment time or lack of confidence are possible reasons (Oliveira et al., 2017).

In other words, the development and diffusion of patient-developed solutions are confined to small groups of peers, and these activities are hidden from the healthcare system and the general population. This is both a problem and a rich potential opportunity.

The concept of 'free innovation' involves building on decades of user-led innovation research and is predicated on a new or different model of the innovation process (von Hippel, 2016). Here the traditional model sees an innovating organization – perhaps an R&D lab, perhaps a small team of clinician experts – developing solutions for a target market of sufficient size to justify the risky exploratory research. By definition, without signals of the likely market potential this research will not happen, effectively closing down many potentially interesting avenues before they have even been started.

By contrast the user-led model, as we see, is about self-interest and immediate peers (Oliveira et al., 2018). The innovation incentive is solving a problem; if there are useful spillovers which others can benefit from then this is a bonus but the original innovators are content with solving their own problem. There is no incentive or desire to move to scale.

The potential of the 'free innovation' model is to bring these two worlds together. Mobilizing the front end risky exploration by capitalizing on user insights, sticky knowledge, and so on, and early prototyping across tight but small peer communities. And linking that to the 'productionizing' and marketing expertise of the mainstream innovation system which is capable of scaling up once initial trajectories are established.

This is a potential win–win collaboration and one which is seeing increasing interest in the commercial sector. Some evidence described in Chapter 4 confirmed that patient innovation can turn out to become a market success once commercialized and scaled up. For example, a shower shirt developed by a breast cancer patient was patented, approved by the FDA as a Class 1 medical device, and commercially available in 36 countries. Here close user engagement especially with pioneer and hero innovators at the extreme end of our spectrum could provide early stage input to innovations which could be developed and scaled for much wider audiences. What is needed are the mechanisms to bring these two worlds closer together.

In our cases there are some examples of hero innovators (Chapter 4) whose concerns and passion drive innovation, sometimes founding their own commercial start-ups, at a minimum provide fuel for 'free innovation'.

There are opportunities here for the healthcare system to become active partners in this process – in the nature of which they are able not only to get improved quality design but also downstream acceptance and diffusion. This is potentially rich territory where taking an inclusive RI approach can pay dividends. The question of how to enable it emerges – and there may well be useful templates to borrow from examples of such partnerships elsewhere in the commercial sector.

Co-creation with active user innovators can take many forms – from Lego's innovative approach to working with its users (mostly children) in designing and scaling new toy ideas), through Local Motors' building of a user community which works in collaboration with large auto manufacturers and component suppliers (Birkinshaw, Bessant and Delbridge, 2007). Many organizations operate versions of innovation labs and co-creation spaces which offer support and facilities to user innovators but also give the host organizations a window on interesting developments and the chance to engage at an early stage with interesting ideas. These co-creation spaces can exist in both online and physical forms and some hybrid collaboration approaches are now emerging in a variety of sectors (Trifilova, Bessant and Alexander, 2016).

15.3.2 Anticipation

This question is essentially asking about the extent to which the downstream implications of innovation are explicitly considered and the ways in which negative outcomes might be anticipated and ameliorated. At the heart of this is a need to make things explicit, articulating the likely shape and direction of the technological changes proposed. It is where tools like technology assessment come to the fore, trying to imagine and simulate alternative possible trajectories and outcomes (Le Feuvre et al., 2016; Inghelbrecht et al., 2016).

In a perfect world this information would emerge and be discussed and explored in an open fashion. In reality there are significant limits to this: RI is not always articulated and because of this it is possible for trajectories to become closed off, for a bandwagon to start rolling. It is much harder to stop once a system is in place than at the start-up stage. The idea of 'technological trajectories' and 'dominant designs' are powerful here, especially in an emerging field like digital healthcare which still has many of the characteristics of what Abernathy and Utterback call the 'fluid' state (Abernathy and Utterback, 1978).

Our examples highlight this – the start-up cases which we observed (Brazilian start up Hand Talk described in Chapter 12, or USA-based start-up WellStart Health, described in Chapter 14, or three Norwegian-based start-ups described in Chapter 10) generally had a higher level and an explicit concern for patient inclusion. Further, Chapter 13 by Naughton and Foss brings the university into the mix. Here we see that the university might be an interesting gate-keeper and contributor to RI through academic freedom and a research focus. By contrast, when innovations are introduced into a context where there is already an established system, such as electronic health records (EHR), it is often pushing against an established base and trajectory, and may run aground on the rocks of too much system change being required. Examples of challenges of that kind are described in Chapter 5 and Chapter 10. Chapter 13 allows for the idea to use the university and its surrounding innovation system to a higher degree.

One characteristic of start-up innovation is an increasing emphasis on tools like the Business Model Canvas (BMC) which explore and identify key stakeholders and their interrelationships (Osterwalder and Pigneur, 2010). This is of value in a context where resources are scarce and uncertainty is high – but the same toolkit may also be of value elsewhere in the innovation process where it offers a helpful framework to explore stakeholder issues and potential inclusion.

Structured frameworks of this kind make explicit not only the original value proposition and delivery mechanisms but also allow for a degree of stress testing – of 'pre-mortem' analysis. These can be helpful in focusing anticipation and exploring potential issues downstream.

A good example can be seen in the use of the BMC in the healthcare TV case (Chapter 7), which identified at an early stage the differences in perspectives and desired outcomes of different stakeholder groups. This led to a 'pivot' in the thinking around the innovation and a split into two parallel projects developing solutions for different groups of stakeholders.

Another approach which helps support the anticipation agenda is the use of tools and approaches linked to the 'agile' or 'lean start-up' (LSU) model for entrepreneurship (Ries, 2011; Cooper, 2017). Here instead of trying to plan an innovation in detail and then implement it, attention is placed on developing and testing hypotheses early through a series of experiments. These probes, using a 'minimum viable product' (MVP) prototype, allow for early user feedback and rapid modification – 'pivoting' – around the original idea. In other words they offer excellent mechanisms for early-stage exploration and anticipation.

The cases described in Chapter 11 (CareConnect), Chapter 12 (Hand Talk and SOSPS), Chapter 14 (WellStart) and Chapter 8 (Laerdal) all provide evidence of the value of this kind of approach.

In the 'Blink' case (Chapter 9) we can see the power of such approaches. Even though there is careful planning and attention to user needs at the front end there is still scope for unexpected issues to emerge. The methodologies in use by the company allowed for an 'early reveal' of problem issues which could then be dealt with in a revised version of the innovation – essentially the above principle of 'pivoting' in practice.

So there is scope for anticipation enabled by tools, by agile methodologies like LSU and indeed by the nature of the technology itself which is reprogrammable rather than fixed in stone. The difficulty emerges when systems are already in place and where they are complex and interdependent. Electronic Health Records (EHR) are a good example in which several generations of IT-based solutions have been developed and implemented. Still, integration is on its way, as demonstrated in Chapter 6 (MyChart case).

Here – as in earlier generations of integrated IT applications in manufacturing – the design space has effectively been constrained by prior actions (Tidd, Bessant and Pavitt, 2005). There is little scope except for tinkering around the edges, softening interfaces or bolting on modules with a more explicit user emphasis. This is the emerging position reported in Chapter 5, where the potential of user engagement is clear but where its implementation may be limited by conflicts with the established system architecture and the underlying culture associated with it.

15.3.3   Reflexivity

This question deals with how far can the innovators question themselves and how flexible are they in using that feedback? As we've seen in the above discussion, the potential for RI is there and the first three dimensions are – to a large extent – visible at last in nascent form. But there is a major challenge implicit in the last, which has to do with context – the social/economic/technological system within which the innovation is located.

In the case of start-ups, the context issues surface because entrepreneurs find their ideas often run up against 'institutional walls' – resistance to change coming because of their status as external to that system. The discussion in Chapter 10 highlights this kind of difficulty. But it is not only institutional walls – there is the difficulty of 'breaking in' to systems which are well embedded in institutional and political structures. As pointed out by Christensen, Waldeck and Fogg (2017), healthcare is an example of a sector that resists disruption, due to strong established systems that limit the whole industry. So most innovations are performed so far in a sustaining manner, not disturbing existing incumbents in their core markets. One approach to challenge this situation is to find alternative markets within which to learn and build a track record for the innovation – but this option is not always available in monolithic healthcare systems like Bismarck or Beveridge models described in Chapter 3.

The challenge is not just institutional; there are also significant barriers associated with the underlying culture in many established healthcare organizations. The traditional role of clinicians, the power structures and the underlying behavioural routines make it difficult to enter and challenge. The start-up entrepreneur's difficulties are not simply technological and economic; they also involve requirements for considerable political expertise to navigate these waters.

The context challenge comes through strongly in the case of EHR where the inertia of the established system – with sunk costs, established practice and the implicit power relations who are already programmed in militate against alternative designs. (There is a parallel to this in the experience of large-scale integrated production management systems (Bessant, Kaplinsky and Lamming, 2003; Fleck, Webster and Williams, 1990). It is also evident in other contextual variables – for example the challenge of start-up entrepreneurs in the university system, as highlighted by Naughton and Foss in Chapter 13. Here the entrepreneur professor clashes with an institutional framework which is pushing its agenda – and where scope for RI may be squeezed out because those institutional policies preclude it.

15.3.4 Responsiveness

The issue here is how far systems are designed to be open and flexible, able to pivot in the light of early feedback. As we have already commented, a big advantage of digital technology is its programmability – and reprogrammability. To this, we should add the potential offered by agile methodologies, which allow for extensive revision during the development phase through a process of prototype test and pivoting. Therefore, the potential for adopting a responsive approach is quite high in the field of digital healthcare.

In practice for many of the more experienced innovating organizations (like Laerdal or Lyse, Chapters 8 and 9), user experience (UX) design is a well-established field and the toolkit for working with users is quite advanced. Therefore, the idea of 'delaying the freeze' and even revising the launched product/service becomes more of an option. And for many of the start-ups explored (Chapters 10, 11, 12, 14) the dominant model of entrepreneurship now embraces with methodologies increasingly influenced by 'agile' thinking based around lean start-up ideas. Such an approach lends considerable force to the potential responsiveness question.

Where this is more difficult is the system change typified by EHR where there is an established trajectory, where the design and implementation process is far less flexible and where – except by 'tweaking' around the edges (the MyChart example) – it is difficult.

Another strand to responsiveness is in the articulation of different viewpoints to which it is necessary to respond. In the case of Health TV, for example, the initial process was inclusive and driven by a responsible agenda. Early stage use of the BMC with the wider steering and supporting group meant that the 'fault lines' began to appear early – as the concept was articulated it became increasingly clear that there were two potential directions which this innovation could take. One solution was unlikely to meet both sets of needs and so the project eventually pivoted to spawn two parallel innovations, both of social value but serving very different 'markets'. Arguably the value of the BMC tool here was as a shared prototyping/articulation framework which enabled the issues to become visible and helped foster responsiveness.

15.4   CONCLUSION

To summarize, responsible innovation matters – of course. But we need to convert it from a 'Motherhood/apple pie' slogan to a practical approach and one whose benefits are clear and which encourages innovators to take this approach. In particular the idea of inclusion of multiple stakeholder views and insights is of interest; whilst it is difficult and time consuming to do so the growing evidence is that engaging with this perspective can enhance design and diffusion. In parallel with the experience of the quality movement back in the 1980s, when the perception of quality management as a necessary but undesirable cost was transformed to a realization that it was in everyone's interest to take on the concern for quality and to build it in at source (Ishikawa, 1985), RI might become a widespread way of understanding the innovation process.

In order to enable RI there is, of course, a need for a framework and guidelines around which to build such innovation. We have been working in this book with the Stilgoe, Owen and Macnaghten (2013) tool, which helpfully identifies four key dimensions and this has helped not only our analysis but could also offer a template for RI design. However, we would argue that taken alone, the framework might be seen as a little prescriptive and static. What is missing is some way of looking at RI as being enacted through a process over time. We propose (and explore in Chapter 2) a model of purpose (the degree to which there is a will to look for RI) and process (the degree to which RI can be operationalized during the innovation journey) should lead to better outcomes more in line with RI.

We can see this time-based evolution of RI capability in several of our cases. For example the Laerdal company (Chapter 8) began with a strong sense of purpose linked to the RI agenda. Founded in response to a crisis event (the near drowning of the founder's son) the subsequent development of enabling products and services to help save lives gives an important social mission to the business. The idea of widespread inclusion of different perspectives was another key foundation; the need for bringing in a wide variety of expertise drew in many medical experts to help develop the first product, the Resusci Anne manikin back in the 1960s. Since then many changes have occurred, but the core principles of the company remained, drawing on two key tributaries, networks of medical knowledge and close user engagement. Today Laerdal is a global company with over 1,500 employees in 24 countries; its core mission remains one dedicated to saving lives through resuscitation, emergency care and patient safety. Embedding RI principles of the kind we have been looking at have helped the business grow while remaining true to these core social values. As the founder Aasmund Laerdal put it:

> ... sustainable business is only possible through developing ability to listen, endless curiosity, practical problem solving, respect for customer, hard work and a passion for continuous improvement. (Tjomsland, 2015, p. 16)

The challenge now comes in trying to extend and replicate these values across an increasingly diverse set of partners as the company moves to a more 'open innovation' set of collaborations with external agencies.

Making RI happen is about managing the contestable nature of innovation – its trajectory is always a product of social shaping forces. For any innovation there is design space. It is wide in the early fluid phases of a technological field but there is also still scope for moving the walls of an established trajectory. But in practice there are multiple obstacles – some of which are more susceptible than others to policy intervention whether at state level, where procurement and reimbursement regimes give the state significant shaping power, or at the enterprise level.

For instance, several chapters demonstrated that disruptive innovations in healthcare are meeting an 'Institutional Wall'. Chapter 4, with Oliveira,

Azevedo and Canhão, shows that patients are innovating, but have difficulty diffusing their innovation. The established healthcare system is not prepared to welcome patient-initiated innovations and policies are needed to create a space where such innovations might be commercialized with help of economic actors. Further, Chapter 10 illustrated that because the healthcare context often is heavily politicized with government involvement and large, powerful companies, it is often difficult for innovating patients and other start-ups within this sector to break into the institutional wall. Chapter 13 showed that the university system may be an important actor in developing innovations and policy interventions can further stimulate such actors. Therefore, we could imagine a new healthcare system where the institutional wall was made more flexible through an innovation system that is geared towards these kinds of start-ups.

Our view on inclusive healthcare model is depicted in the Figure 15.2.

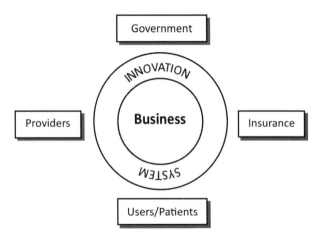

Figure 15.2   Illustrative healthcare model

One area where there is considerable scope for opening up the design space – and a theme on which we have concentrated in the book – is in enhancing the role of users. In the healthcare context this means empowering and engaging patients more extensively. Users are demonstrably a valuable source of insight and as diffusion accelerators and evidence suggests, we should make more of them.

Policy here might include training and empowering patients to bring them further along the involvement spectrum which we have been using in the book.

There might be considerable scope for using procurement policy to help set and shape the direction of such activity, privileging innovations which can demonstrate a high degree of user engagement.

Regional and national initiatives can include the development, as we saw in Chapter 11, of organizational networks, clusters and conglomerates which aim to make user-inclusiveness less time- and resource-demanding for each individual economic actor. There may also be scope for innovation lab-type environments enabling co-creation and facilitation of entry to the mainstream system.

15.5 FUTURE RESEARCH

It is worth concluding with some points which have been raised but not explored in the book. These might well constitute a valuable future research agenda in this important field. We would highlight five key themes:

We have stressed the importance of user engagement and the potential contribution which this offers, but more needs to be done to understand who are the agents who might orchestrate stakeholder's participation?

Typically development of products/services is considered a complex process that requires the management of several factors at different stages, running from concept, through project design and testing to product launch and marketing. A second question relates to improving our understanding of when (at which stage of the innovation process) users might most helpfully contribute?

Innovation can take many forms, for example, product innovation (change in products/services offered by a company), process innovation (change in the way products/services are offered or presented to the consumer), innovation of position (change in the context in which the products/services are introduced in the market) and paradigm innovation (change in the basic mental models that guide the actions of the company). This raises a third question: how can stakeholders' participation contribute to innovation in these different forms?

A fourth question relates to the 'technology' of user participation. As we have seen in the book there are many useful frameworks and methods emerging like the Business Model Canvas. But more needs be done to look systematically at which tools and frameworks are relevant for which stages and for different groups of users.

Finally, we have focused on users and patients in particular. But the RI argument is about widening stakeholder involvement and so there is work to be done around mapping stakeholders and understanding the issues which may emerge in working across a diverse group. These might include conflict of interests, fear of loss of power over the process, fear about the relationship between secrecy and transparency, as well as operational aspects such as time consumption and other resources.

REFERENCES

Abernathy, W. J. & Utterback, J. M. (1978). Patterns of industrial innovation. *Technology Review*, 80(7), 40–47.

Bate, P. & Robert, G. (2007). *Bringing User Experience to Healthcare Improvement: The Concepts, Methods and Practices of Experience-Based Design*. Milton Keynes: Radcliffe Publishing.

Bessant, J. & Maher, L. (2009). Developing radical service innovations in healthcare: the role of design methods. *International Journal of Innovation Management*, 13(4), 555–568.

Bessant, J., Kaplinsky, R. & Lamming, R. (2003). Putting supply chain learning into practice. *International Journal of Operations & Production Management*, 23(2), 167–184.

Bevan, G. & Hood, C. (2006). What's measured is what matters: targets and gaming in the English public health care system. *Public Administration*, 84(3), 517–538.

Birkinshaw, J., Bessant, J. & Delbridge, R. (2007). Finding, forming, and performing: creating networks for discontinuous innovation. *California Management Review*, 49(3), 67–84.

Blaya, J. A., Fraser, H. S. & Holt, B. (2010). E-health technologies show promise in developing countries. *Health Affairs*, 29(2), 244–251.

Canhão, H., Zejnilovic, L. & Oliveira, P. (2017). Revolutionising healthcare by empowering patients to innovate. *European Medical Journal*, 1(1), 31–34.

Christensen, C., Waldeck, A. & Fogg, R. (2017). How disruptive innovation can finally revolutionize healthcare. Innosight Executive Briefing. Retrieved 8 March 2019 from https://www.christenseninstitute.org/wp-content/uploads/2017/05/How-Disruption-Can-Finally-Revolutionize-Healthcare-final.pdf.

Cooper, R. G. (2017). Idea-to-launch gating systems: better, faster, and more agile: leading firms are rethinking and reinventing their idea-to-launch gating systems, adding elements of agile to traditional stage-gate structures to add flexibility and speed while retaining structure. *Research-Technology Management*, 60(1), 48–52.

Fleck, J., Webster, J. & Williams, R. (1990). Dynamics of information technology implementation: a reassessment of paradigms and trajectories of development. *Futures*, 22(6), 618–640.

Inghelbrecht, L., Goeminne, G., van Huylenbroeck, G. & Dessein, J. (2016). When technology is more than instrumental: how ethical concerns in EU agriculture co-evolve with the development of GM crops. *Agriculture and Human Values*, 34(3), 1–15. doi: 10.1007/s10460-016-9742-z.

Ishikawa, D. K. (1985). *What Is Total Quality Control?: The Japanese Way (Business Management)*. London: Prentice Hall Trade.

Kamper, S. J., Maher, C. G. & Mackay, G. (2009). Global rating of change scales: a review of strengths and weaknesses and considerations for design. *Journal of Manual & Manipulative Therapy*, 17(3), 163–170.

Le Feuvre, R. A., Carbonell, P., Currin, A., Dunstan, M., Fellows, D., Jervis, A. J., Rattray, N. J. W., Robinson, C. J., Swainston, N., Vinaixa, M., Williams, A., Yan CBarran, P., Breitling, R., Chen, G. G., Faulon, J. L., Goble, C., Goodacre, R., Kell, D. B., Micklefield, J., Scrutton, N. S., Shapira, P., Takano, E. & Turner, N. J. (2016). Synbiochem: Synthetic Biology Research Centre, Manchester: a UK foundry for fine and speciality chemicals production. *Synthetic and Systems Biotechnology*, 1(4), 271–275.

Oliveira, P., Zejnilovic, L. & Canhão, H. (2017). Challenges and opportunities in developing and sharing solutions by patients and caregivers: the story of a knowledge commons for the patient innovation project. *Governing Medical Knowledge Commons*, 301.

Oliveira, P., Zejnilovic, L., Azevedo, S., Rodrigues, A. M. & Canhão, H. (2018). Peer-adoption and development of health innovations by patients – a national representative study of 6204 citizens, *JMIR Preprints*. 30 July 2018:11726, doi: 10.2196/preprints.11726, http://preprints.jmir.org/preprint/11726.

Osterwalder, A. & Pigneur, Y. (2010). *Business Model Generation: A Handbook for Visionaries, Game Changers, and Challengers*. Chichester: John Wiley & Sons.

Pagliari, C., Sloan, D., Gregor, P., Sullivan, F., Detmer, D., Kahan, J. P. ... & MacGillivray, S. (2005). What is eHealth (4): a scoping exercise to map the field. *Journal of medical Internet Research*, 7(1), e9.

Reber, A. S. (1989). Implicit learning and tacit knowledge. *Journal of Experimental Psychology: General*, 118(3), 219.

Ries, E. (2011). *The Lean Startup: How Constant Innovation Creates Radically Successful Businesses*. London: Portfolio Penguin.

Ronksley, P. E., McKay, J. A., Kobewka, D. M., Mulpuru, S. & Forster, A. J. (2015). Patterns of health care use in a high-cost inpatient population in Ottawa, Ontario: a retrospective observational study. *CMAJ Open*, 3(1), E111.

Sacristán, J. A., Aguarón, A., Avendaño-Solá, C., Garrido, P., Carrión, J., Gutiérrez, A. ... & Flores, A. (2016). Patient involvement in clinical research: why, when, and how. *Patient Preference and Adherence*, 10, 631.

Stahl, B., Obach, M., Yaghmaei, E., Ikonen, V., Chatfield, K. & Brem, A. (2017). The Responsible Research and Innovation (RRI) maturity model: linking theory and practice. *Sustainability*, 9(6), 1036.

Stilgoe, J., Owen, R. & Macnaghten, P. (2013). Developing a framework for responsible innovation. *Research Policy*, 42(9), 1568–1580.

Tidd, J., Bessant, J. & Pavitt, K. (2005). *Managing Innovation Integrating Technological, Market and Organizational Change*. Chichester: John Wiley & Sons.

Tjomsland, N. (2015). *Saving More Lives Together – Vision for 2020* (self-published book by Lærdal Medical AS).

Trifilova, A., Bessant, J. & Alexander, A. (2016). Q&A. How can you teach innovation and entrepreneurship?. *Technology Innovation Management Review*, 6(10), online https://timreview.ca/article/1027.

von Hippel, E. (2005). *Democratizing Innovation*. Cambridge, MA: MIT Press.

von Hippel, E. (2016). *Free Innovation*. Cambridge, MA: MIT Press.

Zejnilovic, L. et al. (2016). Innovations by and for the patients: and how can we integrate them into the future health care system, in Albach, H. et al. (eds), *Boundaryless Hospital*. Berlin: Springer-Verlag, pp. 341–357.

Index

ENDORSEMENTS